Springer Water

Series Editor

The book series Springer Water comprises a broad portfolio of multi- and interdisciplinary scientific books, aiming at researchers, students, and everyone interested in water-related science. The series includes peer-reviewed monographs, edited volumes, textbooks, and conference proceedings. Its volumes combine all kinds of water-related research areas, such as: the movement, distribution and quality of freshwater; water resources; the quality and pollution of water and its influence on health; the water industry including drinking water, wastewater, and desalination services and technologies; water history; as well as water management and the governmental, political, developmental, and ethical aspects of water.

More information about this series at http://www.springer.com/series/13419

Zheenbek E. Kulenbekov · Baktyiar D. Asanov
Editors

Water Resource Management in Central Asia and Afghanistan

Current and Future Environmental and Water Issues

Springer

Editors
Zheenbek E. Kulenbekov
American University of Central Asia
Bishkek, Kyrgyzstan

Baktyiar D. Asanov
American University of Central Asia
Bishkek, Kyrgyzstan

ISSN 2364-6934 ISSN 2364-8198 (electronic)
Springer Water
ISBN 978-3-030-68339-9 ISBN 978-3-030-68337-5 (eBook)
https://doi.org/10.1007/978-3-030-68337-5

This Springer imprint is published by the registered company Springer Nature Switzerland AG
The registered company address is: Gewerbestrasse 11, 6330 Cham, Switzerland

Contents

Part I
Assessment of Available Water Resources at Regional and Basin Scales

Chapter 1
Integrated of Water Resources and Environmental Management in the Transboundary Rivers Basins of Central Asia

P. I. Normatov, I. Sh. Normatov, and Q. Odinaev

Abstract The state of food and water security in the Central Asian region is analyzed. The perspective directions and potential opportunities on achievement of water and food security for the countries of the upper and lower reaches of Transboundary Rivers of the region are shown. The necessity for development and implementation of water-saving technologies of irrigation and selection high productive agricultural crops is pointed.

Keywords Central Asia · Food security · Demography · Disaster · Mountain

1.1 Introduction

The modern stage of human development takes place with conditions of constant challenges caused by global climate change, which makes significant changes in the functioning of the components of the ecosystem. In this regard, all efforts of human thought are towards developing mechanisms to mitigate the negative impacts and consequences of climate change, as well as the adaptation of the entire sphere of activity to its cataclysms.

The mountain ecosystem is more vulnerable and particularly sensitive to climate change compared to the other components of the ecosystem. The current trend in the development of natural phenomena (climate warming, extreme natural situations, etc.) is of particular concern to mountainous countries and this stimulates the adoption of drastic measures to mitigate the effects of climate change. For example, by the Decree of the Government of the Republic of Tajikistan on May 3, 2010, the State Program for the Study and Preservation of the Glaciers of the Republic of Tajikistan for the period 2010–2030 (No. 209) was approved to enable continuous monitoring and the study of glaciers in Tajikistan. The need to approve such a program was justified by the fact that despite the small area of the territory (just over 10% of the total area of Central Asia), Tajikistan has more than 11,000 km^2 of glacial area, occupied by more than 14 thousand glaciers, forming more than 65% of the water

P. I. Normatov (✉) · I. Sh. Normatov · Q. Odinaev
Tajik National University, 17 Rudaki Avenue, Dushanbe 734025, Tajikistan

Z. E. Kulenbekov and B. D. Asanov (eds.), *Water Resource Management in Central Asia and Afghanistan*, Springer Water, https://doi.org/10.1007/978-3-030-68337-5_1

resources for the region. The Pamir is the main glaciation zone of the Republic with a total area of 8500 km^2 [1].

The problems with water and the consequences of global climate change are relevant and are a priority for Central Asia now to stimulate scientific, applied, technical, and economic research. The problem of water availability in various sectors of the economy and population has been felt since the 1960s and 1970s. The manifestations of water scarcity have led to the concept of water security. Since then, the scientific and international attention to the problems of water has increased rapidly. In the last decade, studies have been conducted simultaneously with more than 50 international programs, directly or indirectly aimed at solving water security issues [2].

Due to the increasing pressure of global warming on the functioning components of the ecosystem, a platform of scientists and international communities has been formed and mobilized to develop modern mechanisms for the adaptation and mitigation of climactic cataclysms.

A characteristic of climate change unique to the Central Asian region is the diversity depending on the geographical and orographic features of the area.

1.2 Water and Food Security in Central Asia Region

1.2.1 Water Security

Water resources in the Central Asian region are characterized by uneven spatial distribution. The two main transboundary rivers, the Amu Darya and Syr Darya play an important role in the lives of the more than 80 million people living in the region. The water resources of the region are formed upstream, mainly (80%) in the territories of Tajikistan and Kyrgyzstan. These rivers serve to irrigate the agricultural lands of Kazakhstan, Turkmenistan, and Uzbekistan. With the irrigation of 85% of the agricultural lands of the aforementioned downstream countries, approximately 90% of the annual surface runoff of the Aral Sea basin is used. At the same time, irrigated agriculture provides about 30% of the GDP and employment for more than 60% of the region's population [3].

According to the aforementioned group of international experts, the shortage of fresh water is rapidly increasing as demographic factors increase: in a little less than a hundred years (1900–1995) water consumption has increased 6–7 times, which is twice the growth rate of the Earth's population. The total water withdrawal for 1995 in the world was 3750 km^3/year and water consumption 2280 km^3/year. Taking into account the prospects of economic development, population growth, and climate change, water removal by 2025 can be estimated to approximately 4600–7000 km^3/year. By 2025, water use is expected to increase by 15–35% in developed countries and 200–300% in developing countries [4]. Furthermore, one of the Millennium Goals is to provide the world's population with drinking water, it's achievement will represent the one of the main costs in every country and will amount to 10–30 billion USD per year [2].

In the average year, the total surface runoff resources for the Aral Sea basin are no more than 148.5 km^3/year (116.5 km^3/year is natural river runoff) and about 32.0–33.0 km^3/year is return water. Taking into account nonproductive water losses, these resources do not exceed 125.0–133.0 km^3/year [5]. Natural flow resources in the Aral Sea basin are completely exhausted and the region's economy is developing in conditions of increasing water scarcity. Their total use is already 30–50% more than before in the Syr Darya basin and 10% more than before in the Amu Darya basin [4, 6].

1.2.2 *Food Security*

Organically linked to problems of demography, food security is being addressed in different ways in the upstream and downstream countries of the Transboundary Rivers of Central Asia.

The dynamics of population changes in the Central Asian countries are shown in Fig. 1.1. It follows that by 2050, the population of region will reach more than 88 million people, it will increase by more than 30% when using 2015 as the base year [7].

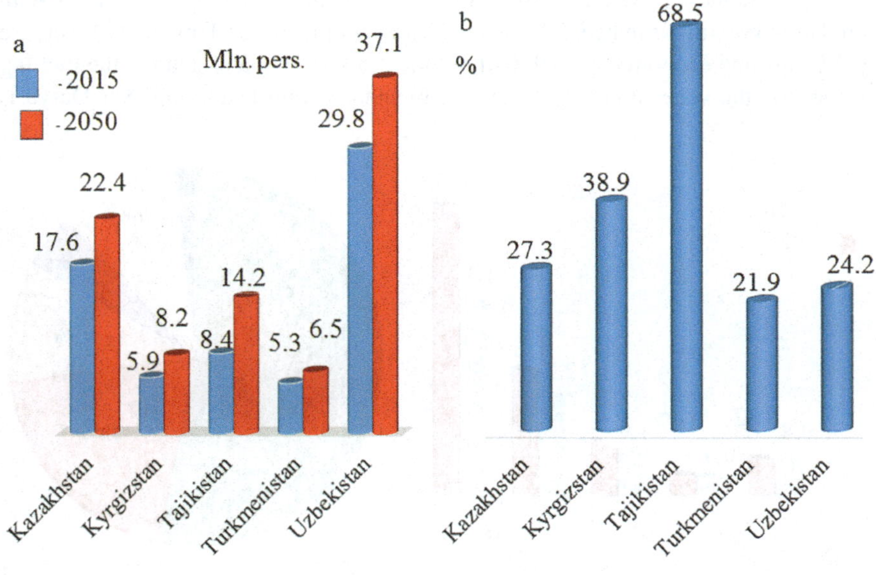

Fig. 1.1 Population increases in Central Asian countries by 2025 versus 2016 **a** and a forecast of the population ratio in 2050—2015 **b**

The water distribution between the countries of the Aral Sea basin is as follows: Uzbekistan withdraws about 53%, Turkmenistan—20%, Tajikistan and Kazakhstan—10%, Kyrgyzstan—less than 5%, and Afghanistan—about 2% [3]. According to the World Bank, about 400 Th. ha of land are currently irrigated in Northern Afghanistan, including directly from the Pyanj and Amu Darya transboundary rivers, about 100 Th. ha. Potential for the growth of irrigated land in the Afghan territory within the basin of these rivers is huge and the area of irrigation of fertile land can reach, according to various estimates, near 500 Th. ha [8].

Consequently, after the stabilization of its political situation, Afghanistan could and most likely will claim an additional volume of water from the water resources of the Amu Darya river basin and the water intake of Afghanistan will reach 3.0 km^3 by the middle of the century. That development will lead to a decrease in downstream flow for other countries.

The downstream countries of the Central Asian transboundary rivers must solve the problem of food security by expanding the area of irrigated agricultural land as the problem is mainly due to sufficient land resources. However, the upstream countries are deprived of such an opportunity due to the limited areas of irrigated land. The total area of irrigated land in the five Central Asian countries in 2016 was 8.7 million ha and by 2025 is expected to increase to 11.8 million ha, an increase of more than 35% (Fig. 1.2).

According to the forecast data of the Scientific Information Center of the Interstate Coordination Water Commission of the Central Asia (SIC ICWC), in the middle of XXI, the water deficit in the utilizing countries of the Amu Darya basin will be 8–11 km^3, including a decrease in runoff from climate change by 1.5–3 km^3. With population growth estimated to be about 21 mln people, water demand will increase by 2.5 km^3 and economic growth will require 1.5 km^3 of water. Due to the melting of glaciers, the water content of the transboundary Amu Darya and Syr Darya is

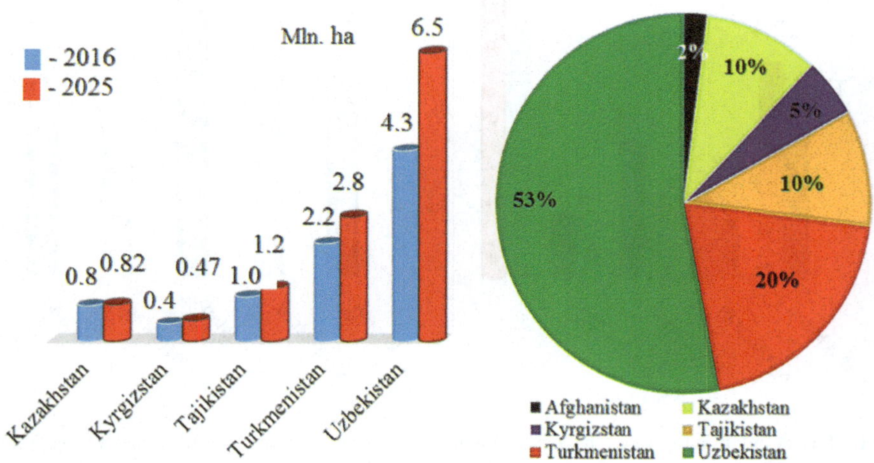

Fig. 1.2 Dynamics of changes in population and irrigated land in Central Asia

predicted to decrease by 12–15% by the middle of the century. For the past 35 years, water supply per capita in the Aral Sea basin has decreased from 4500 m^3 per year to 2150 m^3. Nevertheless, the countries of Central Asia occupy the leading places in the world in terms of water consumption per capita [9].

It follows from the above information that it is time to take decisive measures to ensure the rational usage of water resources and the development of technologies for less moisture-intensive and more climate-resistant varieties of crops. Consistent monitoring of river hydrology and the state of watersheds is necessary for the rational and efficient use of water resources. The meteorological conditions of the basin should also be collected and monitored, thus creating a series of rich and reliable databases, the use of which is the key to making scientifically sound decisions.

1.3 Natural Emergencies and Safety

The problems associated with humans' vulnerability to natural disasters and emergencies in the last few decades have become more and more attention-grabbing. Through a decision made by the United Nations' General Assembly, the World Conference on Disaster Risk Reduction was convened from 18 to 22 January 2005. The purpose of this conference was to identify ways to build the capacities of states and communities to respond to natural disasters and other emergencies and to develop strategic and systematic approaches to emergencies.

According to the Hyogo Framework Program, a World Conference Disaster Risk arises when hydro-meteorological, geological, and other hazards interact with vulnerability factors, such as physical, social, economic, and/or environmental issues. The Hyogo Framework Program is especially relevant for mountainous countries where the influence of meteorological and hydrological cataclysms on various aspects of human life and human habitat is more subtle and sensitive. For example, river floods in the main river basins of Tajikistan occur mainly in the spring due to heavy rains.

According to the information from the Emergency Commission of the Republic of Tajikistan in 2018, more than 169 natural emergencies and accidents were registered. More than 23 emergencies from the total number in 2018 caused considerable material damage to the population and the greater economy of the country, the contribution of each of the natural hazards are shown in Fig. 1.3.

In the Zeravshan river basin, compared to other parts of Tajikistan, emergencies were more frequently associated with water-related factors, namely frequent flooding, mudflows, and avalanches. Emergency cases in the river basin average more than 150 per year, about 7% of the total for the entire country. At the same time, most of the population of the basin (Aini, Matcha, and Penjikent districts) annually suffers great economic losses, and sometimes even human losses [10].

The emergencies in the Zeravshan river basin have rain, snow, and glacial causes, and make up 6.5% of total mudflows for the entire country. The results of the economic damage seen in the respective districts from 2002 to 2005 are presented in Fig. 1.4.

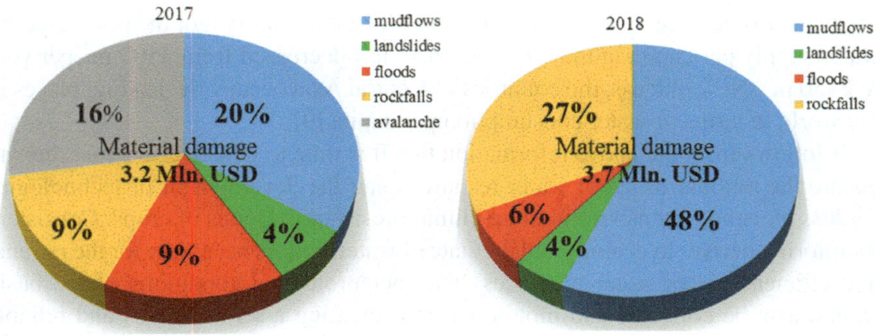

Fig. 1.3 Material damage at various types of natural emergencies

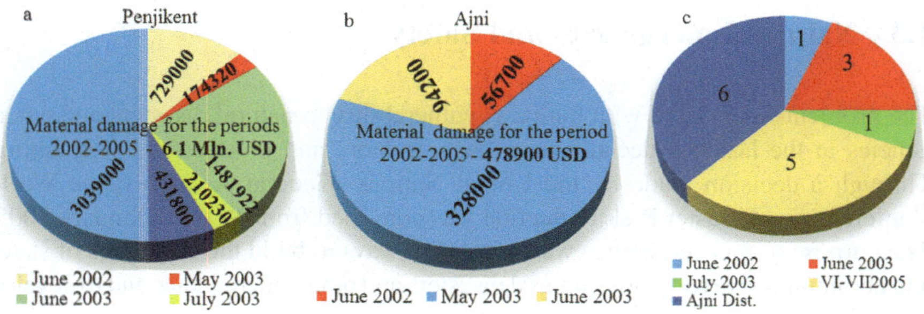

Fig. 1.4 Economic damage (in USD) for floods in the Penjikent **a** and Aini **b** districts of the Zeravshan valley and number of human casualties in the Aini and Penjikent districts for the floods between 2002 and 2005 **c**

The resulting damage to all three main administrative centers, including the Mountain Matcha district of the Zeravshan river are presented in Fig. 1.5. It is seen that compared to the Aini and Penjikent districts, the damage done in the Matcha is relatively small. This is primarily because in the Matcha district precipitation occurs mainly in solid form. In the dispersion zone of the Zeravshan river, in the Penjikent and Aini districts, precipitation mainly falls in the form of heavy rains.

In order to determine the priorities, in terms of developmental direction, and to ensure further sustainable development, the Committee for Emergency Situations and Civil Defense (CESCD) and the Tajik government approved the "Development Program of the CESCD at the Government of the Republic of Tajikistan for 2018–2022" by Decree No. 284 on May 31, 2018. The Development Program of the CESCD will be implemented in two stages, the first from 2018 to 2020 and the second from 2021 to 2022, with the following expenses listed in Fig. 1.6 [11].

The most important conditions for the rapid adoption of measures to protect the population and the territory in the event of a threat. The most important measure to be taken with the occurrence of natural disasters is the availability of timely warnings for the population through the creation of appropriate automated systems. Taking

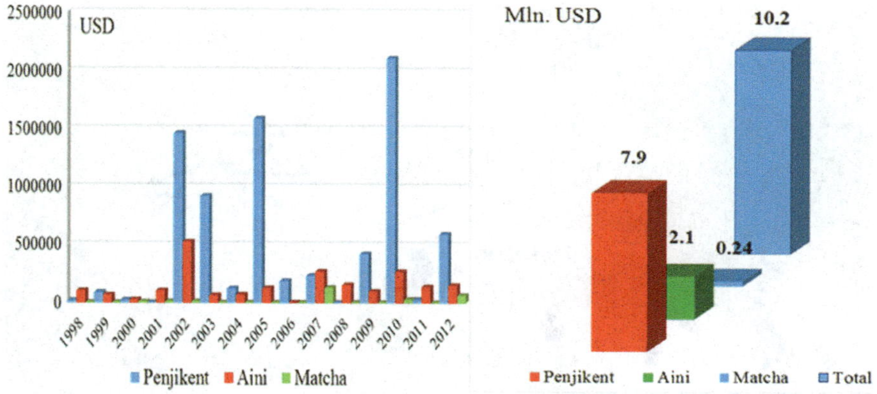

Fig. 1.5 Damage of natural emergencies to three main administrative centers of the Zeravshan River Basin and total volume damage for the periods 1998–2012

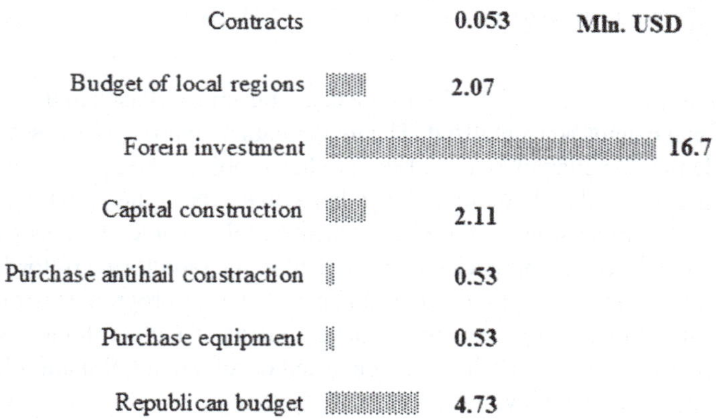

Fig. 1.6 Sources of the expense funds for prevention of natural emergency

into account the importance of the timely informing of the population about possible threats of emergencies, the CESCD installed early warning systems based on the equipment of the GSM IT-70. The GSM system's installation in places of likely mudflows in the mountain areas of Lyakhsh, Rasht, Tajikabad, Nurobod of the Rasht Valley, and Penjikent is shown in Fig. 1.7 [11].

River floods in the basins of the main rivers of Tajikistan occurs mainly in the spring season due to heavy rains. A distinguishing feature of these floods is the sudden transition from floods to debris flows. The transient nature of the floods and the formation of debris flows is mainly due to orographic features. In narrow gorges and steep valleys, incoming air masses are quickly saturated due to favorable thermodynamic conditions and its unloading is accompanied by heavy rains.

Almost all methods of forecasting natural emergencies or disasters are based on heuristic and mathematical approaches. The essence of the heuristic approach is to

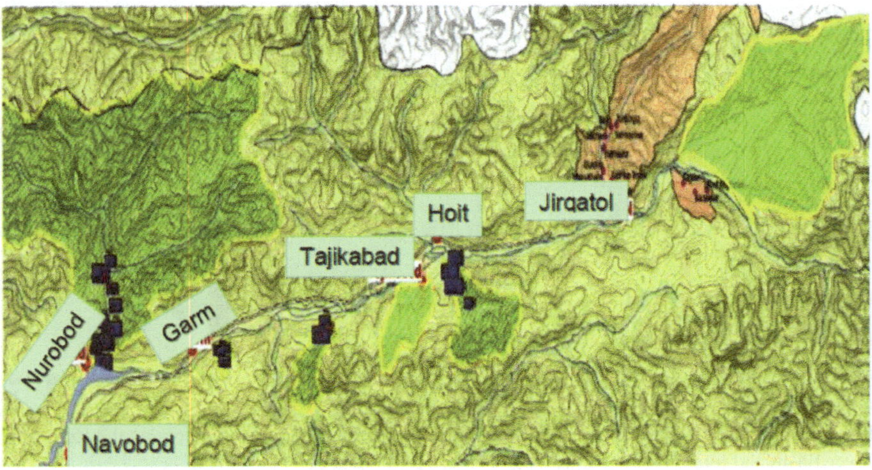

Fig. 1.7 Scheme of early warning systems of CESCD based on the equipment, GSM IT-70. Source: Committee of Emergencies and Civil Defence Republic of Tajikistan

use the opinions of experts in the field. The heuristic approach is then used to predict processes that cannot be formalized. The mathematical approach consists of the use of available data on characteristics of the predicted object, then processing through mathematical methods, obtaining a dependence that relates these characteristics to time, and calculating using the found dependence of the characteristics of the object at a given time. Such an approach can be applied to the monitoring of lowland areas prone to natural disasters, where the influence of terrain orography is minimized. There are difficulties in forecasting the development of meteorological scenarios, such as generating factors for the emergence and development of natural hazards in mountain areas associated with the following factors:

(1) the absence of any dependence of the atmospheric precipitation on the height of the races and the location of the terrain above sea level;
(2) the individuality of each mountain gorge and valley by their meteorological conditions.

In mountainous areas consisting of a set of local territories (foothills, gorges, canyons, etc.) each with their own distinct orography, the climatic conditions presented in the data collected by single meteorological stations are not a reflection of real situations.

1.4 Conclusion

Thus, through the analysis of the state of water usage in the agricultural sector of Central Asian countries, and taking into account that this sector consumes more than

80% of water resources, it was shown that in the condition of climate change and with the intense melting of the region's glaciers, it is necessary to revise current water use norms. It is imperative that the effective and rational use of water resources is primarily dictated by demographic factors. The problems of reducing the areas of glaciation, runoff of river systems, and the negative impact of climate change on ecosystem components, including water resources, requires this.

Through the Zeravshan River basin example, an increase of the natural emergencies accompanied by colossal economic damage and even human losses is clear. It should be noted that the adoption of administrative and economic decisions on the usage of water resources requires preliminary monitoring of the state of water objects. Furthermore, on its basis, the creation of a reliable database, is the key to a scientifically sound approach for solving water problems, mitigating the effects, and perhaps even preventing emergencies.

References

1. Normatov PI, Armstrong R, Normatov ISh (2019) Assessment of the influence of climate change on snow-ice resources and hydrology of Mountain Pamir rivers. J Sustain Dev Mount Terr 11(3):295–304
2. Kotlyakov VM, Seversky IV (2006) Glaciers of Central Asia: current status, changes, possible impact on water resources. In: Assessment of snow-ice and water resources in Asia. UNESCO, Almaty, pp 22–60
3. Medvedev A (2017) The history of the problem of using the transboundary water resources of Central Asia and the role of non-regional partners in resolving them. [Online]. Available at: https://www.j-een.com/news_view/702/. Accessed on 3 Mar 2019
4. Shiklomanov IA (1998) Global renewable water resources. Water a looming crisis? In: International conference on world water resources at the beginning 21st century proceedings. UNESCO, Paris, pp 3–14
5. Severskiy IV, Chervanyov I, Ponomarenko Y et al (2005) Aral Sea. GIWA regional assessment 24. University of Kalmar. Sweden, p 87
6. Kipshakbayev NK, Sokolov VI (2002) Water resources of the Aral Sea basin—formation, distribution, use. In: Water resources of Central Asia. Tashkent: SIC ICWC, Tashkent, pp 63–67
7. FAO (2018) The state of food security and nutrition in Europe and Central Asia. FAO, Budapest. 77
8. Zakhvatov A (2016) With the end of the war in Afghanistan, the struggle for water will intensify in Central Asia. [Online] Available here:https://tj.sputniknews.ru/columnists/20161216/102 1325166/voina-afganistan-borba-voda.html. Accessed on 9 Apr 2019
9. Shulepina N (2018) Water supply in the Aral Sea basin in a changing climate. [Online] Available at: https://nuz.uz/svobodnoe-mnenie/36925-vodosnabzhenie-v-basseyne-aralskogo-morya-v-usloviyah-menyayuschegosya-klimata.html. Accessed on 28 Mar 2019

10. Normatov P, Richard Armstrong R, Normatov IS (2015) Monitoring of emergency water factors and study of anthropogenic impact of industrial facilities on the quality of waterway of the River Zeravshan River Basin. J Meteorol Hydrol 5:89–97
11. Review of emergency and civil defense in the Republic of Tajikistan for 2018. Committee for Emergency Situations and Civil Defense at the Government of the Republic of Tajikistan. Dushanbe (2018)

Chapter 2
Experience of Preparation of River Basin Planning and Management on Water Resource Development, Use, and Protection in Main and Minor Transboundary River Basins

E. P. Sahvaeva

Abstract The Integrated Water Resource Management (IWRM) concept was discussed at large international conferences all over the world in the 1990s and into the beginning of this millennium. As a result, the Special Directive of the World Summit on stable development was adopted in 2002, according to which all countries should have developed Integrated Water Resource Management and Water Saving Plans by 2005. The IWRM's main goals are as follows:

- to prevent the further water deterioration,
- to improve the condition of water ecosystems and related wetlands,
- to facilitate the stable water use based on its long-term protection,
- to ensure decreased groundwater pollution and prevent its further pollution in the future,
- to contribute to flood and drought control.

Keywords IWRM · River basin plans · Ecosystem · Development

2.1 Introduction

Kyrgyzstan committed to IWRM principles by adopting the Water Code in 2005 which is the legal basis for water resource management in the Republic. According to this, the River Basin Plans (RBP) on water development, use, and protection are considered to be the main tool of water management within the hydrological boundaries of the main river basins.

In Kyrgyzstan, work on RBP development started under "Water Management Improvement," a World Bank Project begun at the close of 2007 and is still currently ongoing under the implementation of the "National Water Management Improvement—Phase 1" WB & SDC Project. The first versions of five RBPs

E. P. Sahvaeva (✉)
Kyrgyz Research Irrigation Institute, 68 Mederova str., Bishkek, Kyrgyzstan
e-mail: tadar51@mail.ru

Z. E. Kulenbekov and B. D. Asanov (eds.), *Water Resource Management in Central Asia and Afghanistan*, Springer Water, https://doi.org/10.1007/978-3-030-68337-5_2

Fig. 2.1 Main river basins in Kyrgyz Republic produced by WB & SDC Project [1]

have been prepared for the following locations: the Talas, Chu, Issyk-Kul-Tarim, Naryn-Syrdaria, and Karadaria-Syrdaria basins, as shown in Fig. 2.1.

Besides this, RBPs for the transboundary rivers, the Aspara between Kyrgyzstan and Tajikistan and the Aspara between Kyrgyzstan and Kazakhstan, were developed under the "Transboundary Water Management in Central Asia" Program implemented in the Central Asian region with the support of the German Agency for International Cooperation (GIZ).

In recent years, work on the River Basin Plans for transboundary rivers like the Kurkureusu shared between Kyrgyzstan and Kazakhstan, the Padysha-Ata split between Kyrgyzstan and Uzbekistan, and the Aksu and Isfana rivers, split between Kyrgyzstan and Tajikistan, as well as adjusting the Aspara River Basin Plan, mentioned earlier, has been finalized with support from USAID under their "SMART Waters" Program. Furthermore, the Chon-Aksu River Basin Plan in the Issyk-Kul Lake basin was developed with the support of Norwegian Government under their "Promoting Water-Related Ecosystem Services Concepts in Central Asia" project. Thus, the development of RBPs has been performed with donor support from the WB, SDC, USAID, and Norwegian Government programs.

The main goals of RBP development is the assessment of the following:

- risks of water shortage, drought, floods, pollution and dam failures in a basin;
- the water quantity and quality in a basin, considering expected climate change;
- investment and potential financial needs and their sources.

The main goals of RBP development is also working to identify the following:

- areas under pollution, flood, or mudflow risk;

- current and potential future water needs;
- water reserves with any possible additional uses, considering environmental requirements and international law commitments;
- water demands for environmental needs and population requirements;
- locations where bank protection dams will be constructed for agricultural land protection and strip forests that will be planted;
- areas where gravel and other materials can be mined.

Lastly, the main goals of RBP development are to work to determine the following:

- current and possible water needs by the designation of different purposes;
- water use priorities;
- possible limitations of water user rights within different economic sectors.

RBPs were developed to consider scientific results in different areas, such as hydrology, geology, hydrogeology, glaciology, global climate change, etc. It should be mentioned that River Basin Plans were prepared utilizing the Regional Ecological Center in Central Asia deliverables, particularly the "5 Steps to Basin Planning" and "Guidelines on Basin Planning" publications, and international experience, mainly from the European Union. To develop and implement RBP activities, Basin Councils were set up in main and minor transboundary river basins. While developing RBPs, Basin Councils played an important role along with support given by experts.

The RBP preparation and implementation process consisted of a few main steps. First, the collection and analysis of the information required. Then, identifying any existing problems, developing RBP text, GIS maps and annexes, and developing an activity list identified issues which needed to be addressed and ranking priorities. Later, determining necessary funding, executive bodies and responsible persons to be involved, and discussing the RBP with the Basin Council members. Lastly, the monitoring of the implementation of activities and decisions made.

Besides this, the procedure rules for Basin Plan acceptance should be followed. For instance, a RBP must be developed, adopted, and implemented with the coordination between a Basin Council, consisting of representatives of the state agencies acting within a basin, and the civil sector. River Basin Plans also must be approved by the National Water Council.

As RBP development is a relatively new sphere, some difficulties should be mentioned. For example, all official reporting on water and land use, pastures, etc., is presented by administrative units, not hydrological ones. There is an obvious lack of monitoring for climatic, hydrological, and hydrogeological parameters and no ecological monitoring in Kyrgyzstan, which is important for basin planning.

2.2 Analysis

The following common scheme is usually used to determine the cause-effect analysis of a river basin:

Moving force: social and economic factors and activities increasing or decreasing the load on the environment;

Loading: direct anthropogenic pressure on the environment due to effluents or pollutant discharges and natural resource use;

Condition: current status and trends of the changing environment, including water quality;

Impact: the influence of environmental change on human health and other organisms, also the negative effects on nature and biodiversity;

Reacting: concrete actions focused on solving ecological problems.

The scheme given for Kyrgyzstan is presented in Table 2.1 (shown below) and shows that almost all the national economic activities more or less impact/load on water bodies and in most cases, negatively affect the environmental components.

Table 2.1 Activities and loading on water bodies

Activity	Loading
Housing and utilities sector	Water abstract for household and municipal needs
	Pollution of surface and ground waters by organic, biogenic and synthetic substances
	Contamination by wastewater and decomposition products of municipal solid waste
Power generation, mining and processing industries, transport	Water abstract for industrial purpose
	Disruption of natural hydrological regime of water bodies, deterioration of ecological condition of river systems
	Pollution of air, surface, and ground waters by dangerous substances, oil products
Agriculture	Water abstract for irrigation, significant losses, disruption of river hydrological regime, ecosystem degradation
	Soil erosion, groundwater table rise, settlement, and infrastructure underflooding
	Increased mineralization of surface and ground waters, oil salinization, desertification expanding
	Washing away humous soil layer, increased solid river runoff
	Water pollution by pesticides, organic, and biogenic substances
Fishery	Invasion species, contamination by organic substances
Tourism	Forest cutting, trampling grasses, increased erosion, discharge of household waste into water bodies

Moreover, it does not look possible to assess them in quantitative terms due to a lack of monitoring.

Considering the most significant loadings on water resources, among them an increase in population which should be mentioned as the population of Kyrgyzstan rose from 5.66 to 6.14 million people, or by 8%. The largest percent of which, a 10% increase, was registered in Karadaria-Syrdaria basin. Due to this, a corresponding increase in water consumption for drinking needs is expected and with that also, an increase in the amount of wastewater being discharged into water bodies causing their contamination.

2.3 Conclusion

In recent years, loading on pastures has risen due to an increase in livestock with the largest increase in livestock being in the Talas, Chui, and Karadaria-Syrdaria basins. When grazing is not regulated and grasses on natural pastures are ravaged, soil water-absorbing and moisture-holding capacities are lost, resulting in dusting, packing, and destroying soil structural aggregates. All of which contributes to the soil washing away, causing erosion and water pollution in some cases.

In recent years, water abstracts for irrigation have been quite stable and do not exceed 10.0 km^3/year. But, due to the planned reclamation of newly irrigated land (66,571 ha) and the improving water supply to existing irrigated land (51,085 ha), increasing water abstracts by a factor of 1.0 km^3/year should be expected. Which, in parallel with the predicted decrease in river runoff, due to the global climate change, might cause water availability problems, especially in minor rivers, beyond as soon as the 2020th.

Besides this, increasing loading on water bodies will occur because within the Kyrgyzstan, water protection zones are not in place at water bodies, the environmental releases from reservoirs are not respected, and environmental runoff of water bodies has not been determined, nor identified while preparing the RBPs.

Therefore, the importance of River Basin Plan implementation in the coming years will obviously serve to regulate the identified loads on water bodies. Undoubtedly, it will positively influence the condition of the ecosystem and public health, which is of utmost importance from the perspective of ensuring stable national development.

Reference

1. National water management improvement—phase 1. WB & SDC Project, 2017, Bishkek

Chapter 3
Hydrologic Model for Runoff Simulation of the Kyzyl-Suu River

Zheenbek E. Kulenbekov, Sagynbek Zh. Orunbaev, and A. Zh. Zhumabaev

Abstract Kyzyl-Suu River's runoff model was simulated using Mike Hydro Basin software; CROPWAT software was used in order to calculate evapotranspiration rates. The precipitation and temperature were gathered from the weather stations located in the study area, Daroot-Korgon and Sary-Tash villages, Kyrgyzstan. As a result, the mean river runoff level of summer 2016 is noticeably lower compared to 2015 and 2017. It is because of the low level of precipitation. Relative high river runoff rates in summer 2017 are due to lower evapotranspiration rates and higher amount of precipitation. First time in Kyzyl-Suu River watershed, Kyrgyzstan, the hydrologic model for runoff simulation of the Kyzyl-Suu River was carried.

Keywords Hydrologic model · Kyzyl-Suu · Runoff · Mike Hydro Basin · Simulation

3.1 Introduction

A river is one of the most essential components of the environment. It is an important source of water and nourishment supply. A river is a vital shared natural resource. Moreover, rivers are dynamic fluvial systems. Natural and anthropogenic activities influence to river systems. Human interventions and ecological changes are significant factors that impact hydrological processes. However, the impact of climate change on different aspects of the local hydrology of the Upper Amu Darya catchment is poorly known, and the timing of water availability and quantitative estimates of hydrologic effects of climate change are essential for solving the local water conflicts.

Additionally, a rapid recession of glaciers and their contributions to current runoff could lead to a progressive reduction of the already scarce resources, and therefore, increase the potential for conflict [1]. In such a context, this paper is a contribution

Z. E. Kulenbekov (✉) · S. Zh. Orunbaev · A. Zh. Zhumabaev
American University of Central Asia, 7/6 Aaly Tokombaev Str, Bishkek, Kyrgyzstan
e-mail: kulenbekov_z@auca.kg

Z. E. Kulenbekov and B. D. Asanov (eds.), *Water Resource Management in Central Asia and Afghanistan*, Springer Water, https://doi.org/10.1007/978-3-030-68337-5_3

to the PEER (Cycle 5) NAS USAID project (Integrated Water Resource Management and Strategic Environmental Assessment of Kabul and Amu Darya Rivers), completed in 2019. Thus, it is essential to monitor and evaluate these systems to comprehend their dynamics. This work aims to create a hydrologic model for the runoff simulation of the Kyzyl-Suu River; analyze the simulation for better water resource planning and further forecast the impact of climate change on the river itself.

3.1.1 Study Area

Kyzyl-Suu River flows through the Alay and Chon-Alai valleys (Kyrgyzstan). The sources of the river are glaciers that are located at Trans-Alay Range. The Vakhsh River, also known as "Kyzyl-Suu" in Kyrgyzstan and as "Surkhob" in north-central Tajikistan, originates in the alpine regions of the Pamir Alay. It flows from north-central Kyrgyzstan to the southwest of Tajikistan [2]. The mouth is Vakhsh river 39°16′13.0″N 71°22′51.0″E, where Kyzyl-Suu joins the Muksu river. Biggest settlements along the river are Sary-Tash, Daroot-Korgon, Jekendi, Dombrachy. The map illustrated (Fig. 3.1) shows a study area of this paper.

3.2 Methodology and Data

This work uses the Mike Hydro Basin tool for runoff calculation. It is the software for integrated water resources analysis, planning, and management of river basins. It was developed by Danish Hydraulic Institute (DHI), Denmark [3]. The software is used for analyzing water sharing issues at international and local levels.

Input data are precipitation, temperature, and evapotranspiration. Precipitation and temperature data are acquired from the land-based observation stations. Evapotranspiration is calculated using the CROPWAT software [4] developed by Food

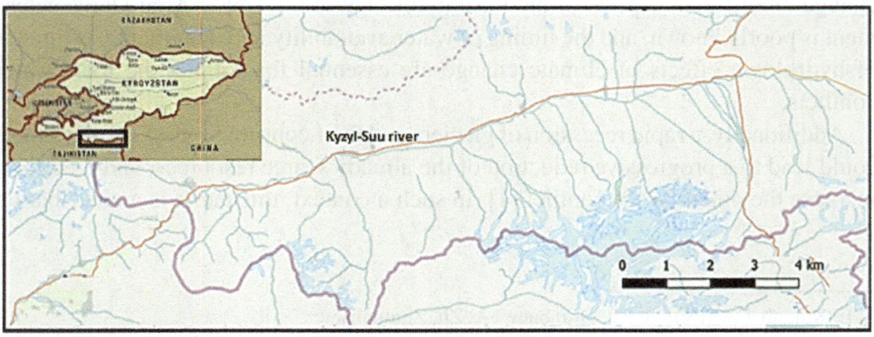

Fig. 3.1 Kyzyl-Suu river basin (Kyrgyzstan, Central Asia)

and Agriculture organization (FAO). Formula used in this software is the FAO Penman–Monteith Eq. (3.1).

$$\begin{aligned}\mathrm{ET}_o = {} & (0.408\Delta(R_n - G) + \gamma 900/(T + 273) \\ & \times u_2(e_s - e_a))/(\Delta + \gamma(1 + 0.34u_2)) \\ & \text{FAO Penman–Monteith equation}\end{aligned} \tag{3.1}$$

where

ET_o	reference evapotranspiration (mm day^{-1}),
R_n	net radiation at the crop surface (MJ m^{-2} day^{-1}),
G	soil heat flux density (MJ m^{-2} day^{-1}),
T	mean daily air temperature at 2 m height (°C),
u_2	wind speed at 2 m height (m s^{-1}),
e_s	saturation vapour pressure (kPa),
e_a	actual vapour pressure (kPa),
$e_\mathrm{s} - e_\mathrm{a}$	saturation vapour pressure deficit (kPa),
D	slope vapour pressure curve (kPa °C^{-1}),
g	psychrometric constant (kPa °C^{-1}).

This formula was developed by experts from FAO, International Commission for Irrigation and Drainage (ICID), and the World Meteorological Organization (WMO) [4].

In order to compute the evapotranspiration values, the software needs to have climatological records of sunshine (hours), air minimum and maximum temperature, humidity (%) and wind speed (km/day). The outputs are sunshine radiation (MJ/m^2/day) and evapotranspiration (mm/day). The results of these calculations can be seen in the graph (Fig. 3.2).

Temperature and precipitation data were acquired from a land-based meteorological observation station located in the Kyzyl-Suu River basin in the Chon-Alay Valley.

The following graph (Fig. 3.3) shows the mean temperature from 1/6/2014 to 30/5/2018 in the Chon-Alay Valley. This data was calculated from minimum and maximum temperature that was obtained from a land-based observation station.

The same timeline presented in the step-accumulated precipitation (mm/day) chart (Fig. 3.4). This barchart illustrates data for each day. Data collection has been done instrumentally.

3.3 Discussion

The programme outputs the runoff data of Kyzyl-Suu River in the Chon-Alay Valley (Fig. 3.5). Results show that runoff values fundamentally depend on seasons; in winter a few waters are drained away because it is frozen due to low temperatures.

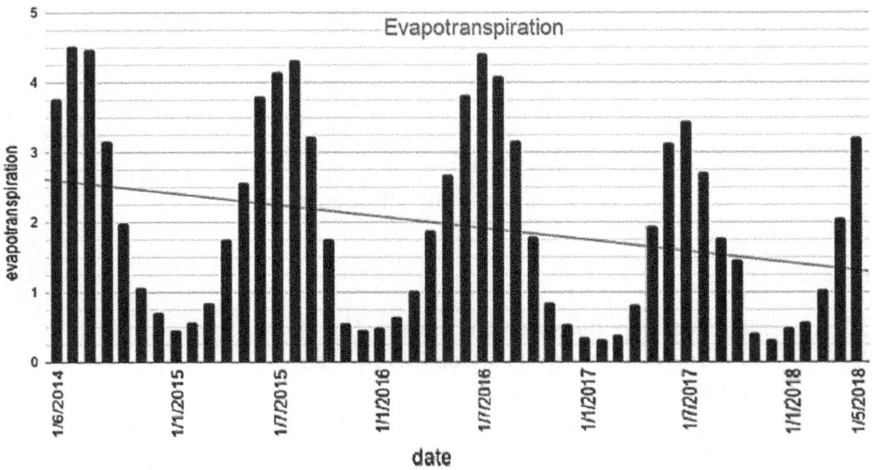

Fig. 3.2 Evapotranspiration (mm/day) data from 2014 to 2018 in the Chon-Alay Valley

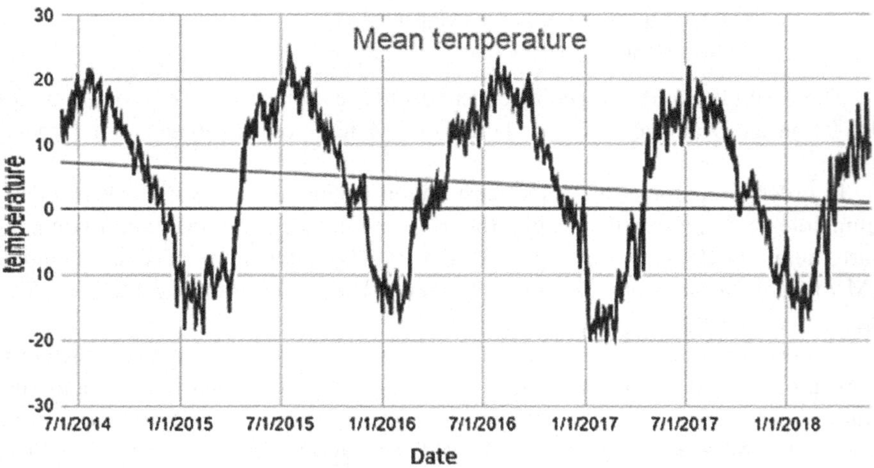

Fig. 3.3 Mean temperature (°C) data from 2014 to 2018 in the Chon-Alay Valley

That produces natural icy and snowy water reservoirs. Relative high evapotranspiration caused by high mean temperature indicates extremely low runoff values in summer 2014. However, it does not correlate with precipitation data in the same timeline. Possibly, software has made a mistake because of the absence of the data before summer 2014.

Mean runoff level of summer 2016 is noticeably lower compared to 2015 and 2017. It is because of the low level of precipitation. Relative high runoff rates in summer 2017 are due to lower evapotranspiration rates and higher amount of precipitation. From the results of these two summers, it can be inferred that precipitation is the most important factor in the runoff calculations.

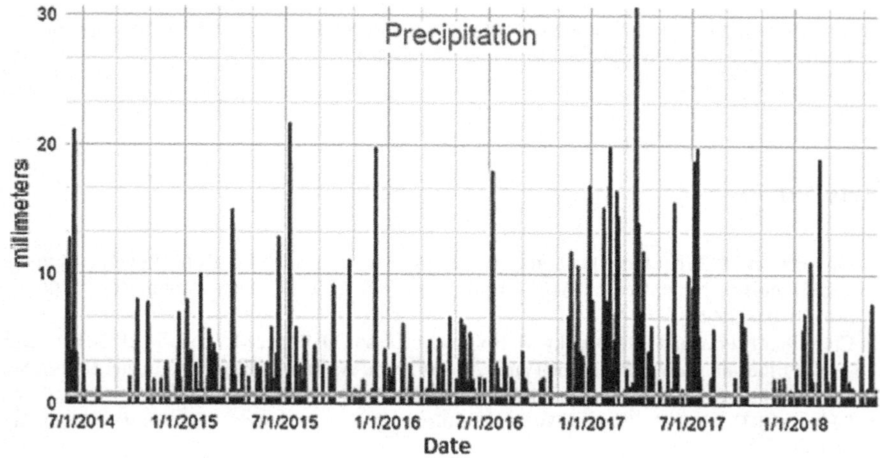

Fig. 3.4 Precipitation (mm/day) data from 2014 to 2018 in the Chon-Alay Valley

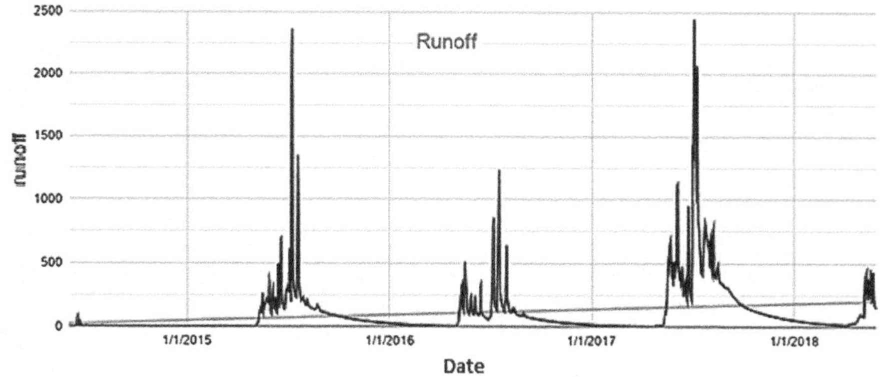

Fig. 3.5 Runoff (m^3/s) data from 2014 to 2018 of Kyzyl-Suu River in the Chon-Alay Valley

3.4 Conclusion

This hydrological simulation model demonstrates that any small changes in environment such as temperature, precipitation, and evapotranspiration have observable impact on the runoff of the fluvial system. Despite the fact that the mean temperature tends to decrease over time, runoff trend line is going up (Fig. 3.3) in the Chon-Alay Valley; which can indicate that glaciers ablation increased overtime or runoff mainly depends on precipitation factor. The study shows that more detailed hydrological simulation is needed and the time range must be increased. The covered timeline is not enough for forecasting, climate analysis, and global hydrological investigation. Further research will be conducted with more data and more variables.

Acknowledgements The research is carried out as a collaborative study and was accomplished by the financial support of USAID and PEER regional project.

References

1. Olsson O et al (2010) Identification of the effective water availability from streamflows in the Zerafshan river basin, Central Asia. J Hydrol 390(3–4):190–197. https://doi.org/10.1016/j.jhydrol.2010.06.042
2. Chevallier P, Pouyaud B, Mojaïsky M, Bolgov M, Olsson O, Bauer M, Froebrich J (2014) River flow regime and snow cover of the Pamir Alay (Central Asia) in a changing climate. Hydrol Sci J 59(8):1491–1506. https://doi.org/10.1080/02626667.2013.838004
3. DHI—MIKE HYDRO Basin [Computer Software]. Danish Hydraulic Institute, Hørsholm, Online. 2020
4. Smith M (1992) CROPWAT: a computer program for irrigation planning and management. FAO land and water development division food and agriculture organization of the United Nations, Rome, p 126

Part II
Impact of Climate Change and Human Activities on Water Resources

Chapter 4
Assessment of Amu Darya Runoff Changes as a Result of Predicted Climate Change and Reduced Glaciation

M. V. Bolgov

Abstract The task of assessing the water resources of the Central Asian Rivers and their forecast under the influence of economic activity and climate change is very important in modern conditions. Water scarcity is the main factor limiting the development of states' economies and it negatively affects the state of water ecosystems. The results of hydrometeorological monitoring show that currently in the region there are noticeable changes in various components of the climate system, specifically; there is a tendency to warming. Observational data in the Amu Darya river basin and model calculations show a steady decrease in carry-over snow reserves and degradation of glaciers. The task of forecasting changes in the water resources of the Amu Darya River is due to possible climate changes and it is becoming urgent. The value of such estimates will increase every year due to a variety of economic reasons, for example, due to high specific water consumption. To solve the problem of assessing and forecasting the water resources of the mountain regions of Central Asia, an approach is proposed where main provisions include: (1) development of stochastic models of meteorological factors for hydrological calculations and forecasts in various climatic scenarios; (2) statistical simulation of a long-term series of meteorological characteristics in various scenic climatic conditions and calculations of river flows in the Amu Darya basin, taking into account the modern glaciation of the catchment; (3) statistical processing of simulation results and determination of calculated hydrological characteristics based on Bayesian ideology.

Keywords Icing · River runoff · Climate change · Stochastic models · Bayesian approach · Hydrological forecasts · Central Asia

M. V. Bolgov (✉)
Institute of Water Problems, Gubkina 3, Moscow, Russia

Z. E. Kulenbekov and B. D. Asanov (eds.), *Water Resource Management in Central Asia and Afghanistan*, Springer Water, https://doi.org/10.1007/978-3-030-68337-5_4

4.1 Introduction

4.1.1 A Subsection Sample

Modern climate change has naturally and anthropogenic-determined components. Since the accuracy of the models of global atmospheric circulation is insufficient, the uncertainty of long-term forecasts is still very high. At present, it is only with a certain degree of probability that it can be argued that the anthropogenic component of warming prevails.

The use of calculation results for global circulation models with an increased (scenario) CO_2 content in the atmosphere is possible, but it should be noted that regional climatic characteristics are reproduced by general circulation models much worse than global parameters. Is it possible to transfer model conclusions about global climate change to the regional level? This is a complex issue that requires taking into account many factors of model concept uncertainty.

From the point of view of hydrological applications, the most important question arises—how to take into account the uncertainty of climate change in hydrological forecasting. Predictive estimates can be obtained by using the Bayesian approach, considering the chances of implementation a particular scenario. The main difficulty is assessing the likelihood of any of these scenarios occurring. If the scenarios adopted by the scientific community do not have obvious preferences over each other, then as a first approximation, it can be accepted that their chances to be realized in the future are equal. The differences between the scenarios can then be considered. For example, introducing a distribution of verification errors or other modeling errors.

In modern conditions and as the value of these forecasts will increase every year, it is very important to assess how the water resources of the Amu Darya will change due to possible climate change. The study of climate dynamics has showed that presently there are changes in the various components of the climate system within the region. Specifically, there is a clear tendency towards warming [1, 2]. Observational data in the Amu Darya River basin and modeled calculations show a steady decrease in carry-over snow reserves and degradation of glaciers.

The problem of the water resources assessment under conditions of significant uncertainty of climate change forecasts, the state of catchments is considered on the example of the Amu Darya river basin and the method for solving these aforementioned issues is proposed. The main provisions of the method under consideration include: (1) the development of a method for stochastic modeling of meteorological factors (in mountain conditions these are precipitation and temperature) for a set of meteorological stations used for hydrological forecasts in various climatic scenarios; (2) imitation of a long series of meteorological characteristics for a set of weather stations in various scenic climatic conditions and the corresponding calculations of river flows in the Amu Darya basin, according to the model of its formation and taking into account modern glaciation of the catchment; (3) statistical processing of the results of simulation experiments and determination of calculated hydrological characteristics based on Bayesian ideology.

To assess the impacts of climate change on the Amu Darya river runoff from its formation zone, researchers from the Hydrometeorological Institute in Tashkent, Uzbekistan built climate scenarios based on models of general atmospheric and ocean circulation (GCMs) using MAGICC/SCENGEN 4.1 [3, 4], where the GCM results are converted to nodes of a single latitude—longitude grid.

4.2 Simulation of Artificial Runoff Sequences, Air Temperature, and Precipitation in the Runoff Formation Zone of the Amu Darya River

4.2.1 Simulation for Water Management Calculations in the River Basin

Amu Darya needs long sequences of flow (tributaries) in the main nodes of the design scheme. The length of rows of inflows should provide the required reliability of water management calculations. The duration of the tributary rows should be at least several decades, which is a difficult condition for the Upper Amu Darya basin. For the Afghan part of the catchment, there are only fragmentary observational data (10–15 years), which forces us to resort to the use of recovery algorithms and simulation of the artificial runoff series. To simulate artificial annual runoff sequences and meteorological elements, an algorithm based on field decomposition into natural orthogonal functions (EOF) is recommended.

Possessing sequences of precipitation and temperatures, as well as information about the state of glaciation (current and forecast estimates), enables the usage of flow formation models. Through these, we calculate the water content of the river and obtain a hydrological series of the required length. The approach allows taking into account possible climatic changes and forecast estimates of glaciation by adjusting the parameters of the hydrological model.

A modeling system developed and verified at NIGMI (Uzbekistan) under the leadership of Denisov Yu. M. For the formation zone of the Amu Darya River flow, it uses information about precipitation collected from 24 weather stations and approximate temperatures collected from 17 weather stations.

4.2.2 Stations Used in Runoff Simulation

Statistical processing of observed data for a representative period was performed, and the next characteristics where calculated: (1) distribution parameters of precipitation and average monthly air temperatures; (2) matrices of pair correlation coefficients for both monthly precipitation and the monthly average temperatures; (3) autocorrelation functions of both precipitation and temperature.

Table 4.1 Fragment of the correlation matrix of the field of average monthly air temperatures

Meteo-station	1	2	3	4	5	6
1 Garm	1.00	0.65	0.85	0.77	0.79	0.86
2 Gulcha		1.00	0.59	0.56	0.63	0.56
3 Gushari			1.00	0.77	0.85	0082
4 Hovaling				1.00	0.75	0.77
5 Fedchenko					1.00	0.77
6 Tavildara						1.00

Table 4.2 Fragment of the correlation matrix of the field of monthly average precipitation

Meteo-station	1	2	3	4	5	6
1 Garm	1.00	0.50	0.72	0.63	0.54	0.77
2 Gulcha		1.00	0.42	0.40	0.56	0.46
3 Gushari			1.00	0.66	0.45	0.64
4 Hovaling				1.00	0.43	0.68
5 Fedchenko					1.00	0.52
6 Tavildara						1.00

Immediate estimates of the distribution parameters showed that the modeling of the average monthly temperatures the normal distribution law was acceptable, and for precipitation—a three-parameter gamma distribution.

The analysis of autocorrelation functions showed that as a model of the time series, one can choose the first-order autoregression process (both for precipitation and temperature) with a remote seasonal variation. In particular, the autocorrelation coefficient of the monthly precipitation sequences totaled $r\ (1) = 0.1$, and for average monthly temperatures $r\ (1) = 0.3$.

Tables 4.1 and 4.2 show fragments of the correlation matrices, the fields of average monthly temperatures, and monthly precipitation for six meteorological stations. The described algorithm was used to simulate artificial realizations of precipitation fields and temperatures under climatic changes. The correlation structure of the fields remained the same as in the expected modern climate, but the parameters of one-dimensional distributions changed in accordance with the results of predictive calculations obtained on models of general atmospheric circulation.

4.3 Bayesian Estimates of Runoff Characteristics

The use of Bayesian methods is caused by the need to take into account significantly different behavior of the river flow process for various scenario conditions.

The probability distribution of river runoff is usually have a positive asymmetry and is described by a two or three-parameter gamma distribution. In this study, we restrict ourselves to a two-parameter gamma distribution as shown below

$$P(x_0, \gamma, x) = \frac{\gamma^{\gamma}}{\Gamma(\gamma)}\left(\frac{x}{x_0}\right)^{\gamma-1} \mathrm{e}^{-\gamma \frac{x}{x_0}} \tag{4.1}$$

where x_0 is the average value, $\gamma = 1/C_v{}^2$, C_v is the coefficient of variation, and $G(\gamma)$ is the Gamma function.

The next step is to calculate the forecast distribution function of the runoff based on the total probability formula. In this case, the predicted density is estimated as follows:

$$\pi(x) = \int\limits_{\theta} P(x, \theta) \cdot p(\theta/x)\mathrm{d}\theta \tag{4.2}$$

the posterior probability distribution of the parameter is θ, and the two parametric Gamma distribution with the known parameter is γ. As a result of numerical integration (2), we obtain the Bayesian predicted probability distribution of the stock characteristic under study.

In our case, three states of the predicted process will be considered below, and we can assume that a system with probabilities n_1/N, n_2/N, and n_3/N can be in one of them. Moreover, $n_1 + n_2 + n_3 = N$, where N is the sum of all probabilities. In this case, the distribution of the mean estimate will be a combination of the three following distributions:

$$\tilde{p}(\theta/x) = \frac{n_1}{N} \cdot \eta_1(\theta, x) + \frac{n_2}{N}\eta_2(\theta/x) + \frac{n_3}{N}\eta_3(\theta/x) \tag{4.3}$$

where the sample distribution of the average value for the i-th scenario with a weight ni is the posterior distribution of the estimate θ for the predicted climatic conditions.

The final stage of the Bayesian prediction is calculation of the forecast distribution which based on the formula of total probability. In this case, the combination of distributions (4.3) will be a priory density, and the model distribution P will be a two-parameter Gamma distribution (4.1) with parameter γ assumed to be the same for all scenarios:

$$\pi(y) = \int\limits_{\theta} P(y, \gamma, \theta) \cdot \tilde{p}(\theta/x)\mathrm{d}\theta \tag{4.4}$$

The resulting distribution predicted will no longer be the Gamma distribution and is obtained by numerically integrating Eq. (4.4).

As already noted, in the Central Asian region, climatologists use six scenarios of future climate changes as the most realistic ones [5, 6]. Two of these scenarios were used in this project and since they lead to the worst hydrological results compared to the modern climate, each of them is given a weight of 1/6. The remaining scenarios led to water values that do not differ much from those typical of the modern climate, therefore, the distribution parameters calculated for a representative period are given a weight of 4/6. In Tables 4.3 and 4.4, you can see the results of calculations by Formula (4.4) for two catchments: the Vakhsh river—p. Komsomolabad and the river Amu Darya—v. Kerky.

4.4 Conclusion

The problem of obtaining forecast estimates for river flow with conditions of significant uncertainty in the factors of its formation is solved by the example of the Amu Darya river basin.

The observational data gathered from meteorological and hydrological stations and posts in the basin were examined and processed. Using the results of calculations on models of general atmospheric circulation and forecasts of the degree of glaciation in the mountainous part of the basin, a set of scenarios of changes in river water content is obtained.

The forecast curve of the annual river runoff in the Amu Darya basin was obtained as a result of the application of the Bayesian forecasting method. The Bayesian forecast flow curve is based on assigning different weights to the estimates of the average for different climate scenarios.

The estimates obtained show a reduction in water resources in the Amu Darya basin.

Table 4.3 The distribution function of the annual runoff of the Vakhsh River, v. Komsomolabad

#	Time	Weight, % for scenarios			Exceedance probability, %				
		Modern climate	Scenario A1	Scenario B1	50	75	90	95	97
1	30 years	50	25	25	560	510	468	444	429
2	50 years	50	25	25	539	479	431	407	392

Table 4.4 The distribution function of the annual runoff of the Amu Darya River, v. Kerki

#	Time	Weight, % for scenarios			Exceedance probability, %				
		Modern climate	Scenario A1	Scenario B1	50	75	90	95	97
1	30 years	50	25	25	1860	1730	1610	1550	1500
2	50 years	50	25	25	1800	1644	1516	1440	1400

References

1. IPCC (2013) Climate Change 2013: the physical science basis. In: Stocker TF, Qin D, Plattner G-K, Tignor M, Allen SK, Boschung J, Nauels A, Xia Y, Bex V, Midgley PM (eds) Contribution of working group i to the fifth assessment report of the intergovernmental panel on climate change. Cambridge University Press, Cambridge, UK, p 1535
2. Parry M (2002) Scenarios for climate impact and adaptation assessment. Global Environ Change 12(2):149–143
3. Using a Climate Scenario Generator for Vulnerability and Adaptation Assessments: MAGICC and SCENGEN. Version 2.4, Workbook, Climatic Research Unit, Norwich, UK, May 2000, p 52
4. Wigley TML (2003) MAGICC/SCENGEN 4.1: Technical Manual, National Center for Atmospheric Research, Colorado, USA, Oct 2003
5. Spektorman TY (2002) Methodology for constructing a scenario of climate change in Uzbekistan using the concept of "ideal forecast". In: Assessment of climate change in the territory of the Republic of Uzbekistan, development of guidelines for assessing environmental vulnerability. Bulletin No. 5., Tashkent: SANIGMI, S. 83–88
6. Spektorman TY, Petrova YV (2007) Climatic scenarios for the territory of Uzbekistan. Climatic scenarios, climate change impact assessment. Bulletin No. 6. Tashkent: NIGMI, pp 14–21

Chapter 5
Climate Change Forecast in Kyrgyzstan for 2050 and 2100

D. T. Chontoev and L. V. Bazhanova

Abstract The article presents the results of the analysis of climate change in Kyrgyzstan, according to observations at sensitive weather stations of the northern and southern regions. Based on a retrospective analysis of temperature changes over the observational period and based on the dynamics (rates) of temperature growth and the preservation of this trend, a forecasted temperature increase for the period up to 2050 and 2100 was reported.

Keywords Climate change · Annual temperature · Annual precipitation · Transboundary rivers · Forecast · Kyrgyzstan

5.1 Introduction

The article presents the results of an analysis of the dynamics of climate change in the regions of Kyrgyzstan and a forecast for the period up to 2050 and 2100, according to observations at weather stations of transboundary pilot basins. This work was carried out as part of the CAREC project, "Water, Education and Science." This project includes a small basin of transboundary rivers, the flow of which is used by two or three states.

Over the past 40 years, climate change has shifted from the topic of scientific discussions to a real problem requiring practical implementations in the modern period. This means forecasts for the future and the development of measures to adapt to changing the environmental conditions.

In Central Asia, climate change will affect river water availability, many of which, both large and small, are transboundary.

Our findings present the results of the analysis of the dynamics of changes in the average annual temperature and precipitation at weather stations of the considered transboundary basins for different time periods of observation. Based on the results

D. T. Chontoev (✉) · L. V. Bazhanova
Institute of Water Problems and Hydropower, National Academy of Sciences of the Kyrgyz Republic, 533, Frunze st, Bishkek, Kyrgyz Republic

Z. E. Kulenbekov and B. D. Asanov (eds.), *Water Resource Management in Central Asia and Afghanistan*, Springer Water, https://doi.org/10.1007/978-3-030-68337-5_5

of the trend analysis and its extrapolation for the forecast period to 2050 and 2100. The forecast-consultation of changes in air temperature and precipitation is given.

This paper used meteorological data from observations of Kyrgyzhydromet [1], as well as published sources (reference books, monographs, articles, reports) on this topic.

5.2 Global and Regional Climate Change

Climate change is a gradual change in the average characteristics that define climate. It is determined by a shift in the following factors: temperature, precipitation, evaporation, etc., over a long period of time. The problem of climate change has become relevant now and will remain so for the foreseeable future, both globally and regionally.

5.2.1 Global Climate Change

The problem of climate change has always attracted the attention of scientists, but in 1970s, it acquired a global character. It was in the 70s that signs of climate change began to manifest themselves more clearly. The expression of these changes was with frequent droughts, catastrophic floods, and an increase in the amplitude of interannual, inter-seasonal fluctuations in air temperature with a general tendency to increase both seasonal and annual average monthly temperatures.

The research results, summarized in the report of the Intergovernmental Group of Experts (IPCC AR5, 2013–2014), estimated that the global average annual air temperature has increased over the past century by 0.3–0.6 °C and the level of the world's oceans have risen by 10–20 cm. The trend of climate warming persists, and especially significant warming, has occurred since 2000. Presumably, by the middle or the end of the current century, the concentration of CO_2 in the atmosphere will double, and the resulting rate of change in the average annual temperature will increase about 0.2–0.4 °C over 10 years, about 2–4 °C per century [1]. The magnitude of the increase in air temperature is consistent with the forecast of global climate warming, due to an increase in carbon dioxide in the atmosphere, a forecast first made by M. I. Budyko more than 40 years ago based on the energy-balance model [2, 3].

It is obvious that changes in climatic conditions will lead to changes in the water resources of river basins, regions, and countries. The general strategy of water supply to the population and economies of the countries in the twenty-first century will also depend on the direction of these changes.

5.2.2 *Regional Climate Change*

Regional climate change has also been observed in Kyrgyzstan. Calculation and analysis of the main climatic parameters, air temperature, and precipitation, was carried out at weather stations in the southern and northern regions located in the pilot basins of transboundary rivers, at different heights, and over the course of a long observational period.

The northern region includes the Chui and Talas basins, which include two pilot river basins, the Aspara and Kurkure-Suu. In this region, the weather stations have the longest observational periods:

Bishkek—1932–2017 (85 years), height 756 m—plain zone;
Baitik—1914–2017 (103 years), height 1579 m—foothill;
Tuya-Ashuu (southern)—1953–2017, (65 years), height 3120 m—high mountain;
Talas—1930–2017 (88 years), height 1217 m, foothill zone [4].

Air temperature

The trend analysis of the average annual temperature at weather stations in the Chui basin (Bishkek, Baitik, Tuya-Ashuu) for the period of parallel observations (1959–2017) is unambiguously positive, an increase of 1–2 °C on an average is noted (Fig. 5.1).

The temperature increases over the course of 59 years. According to the initial and final value of the trends, amounted to 1.90 °C for Bishkek MS, 0.60 °C for Baitik, and 1.00 °C for Tuya-Ashuu, and average growth rate from 0.01 to 0.030 °C per year. The most significant positive dynamics of temperature growth was noted after 1972. Figure 5.2 shows the trends in average annual temperature comparing two periods, one being, 1932–1972, and the other, 1973–2017. The temperature increase from the initial and final trend values amounted to 0.20 and 1.60 °C, respectively. Figure 5.3 and the table accompanying it show the temperature change over 10-year periods, from which it can be seen that the most dynamically increasing temperature began after the 1970s, and the most significantly, after 2000. Thus, the trend of climate warming persists and becomes more dynamic.

According to Bishkek MS (lowland zone), the increase in average monthly temperature in all months is positive with a range of 0.1–2.40 °C. Furthermore, it has become higher than the previous period from 0.1 (May) to 2.40 °C (November). According to MS Baitik (foothill zone), October, February, and March became somewhat colder, with temperature changes of 0.1, 0.2, 0.40 °C, respectively. There was an increase the remaining months from 0.1 to 1.50 °C.

When comparing temperature changes for the Tuya-Ashuu MS (high-altitude) and Zhangi-Zher (valley zone), the following trend was noted: the average annual temperature increased for the Tuya-Ashuu MS by 0.3 °C, for the Zhangi-Zher MS by 0.7 °C, while an increase in the average monthly temperature for this station was observed in all months from 0.3 to 1.80 °C. For Tyu-Ash, a slight decrease in temperature was noted in January–March by 0.3–0.40 °C, and in all other months,

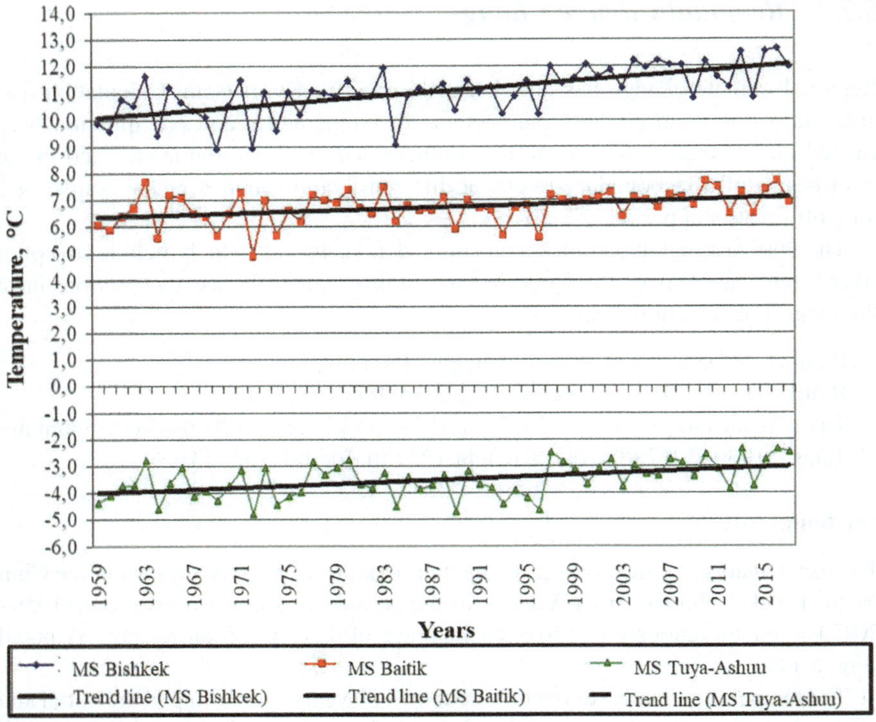

Fig. 5.1 Trend of average annual temperature according to the MS of the northern region for the parallel observation period of 1959–2017

an increase of 0.1–1.0 °C. The increase in air temperature in the valley part is more significant, as the temperature trend decreases with height.

MS Talas also noted a more dynamic increase in temperature after 1972. Trends for the compared periods are presented in Fig. 5.4.

Figure 5.5 shows the temperature increase according to MS Talas for the period between 2012 and 2017, relative to the long-term average value. It is from this data, it can be concluded that the trend of climate warming in recent years has not only persisted, but has also grown more significant over the course of the entire observational period. Table 5.1 shows the results of a comparative analysis of the average temperature and its growth rate for different periods, relative to the norm and the previous period.

From the data presented in the table, we can deduce that the trend of increasing air temperature persists and the most significant years within the observational period was and is the 2001–2017 date range, with an increase of 0.04 °C per year. The same rate of temperature increase was noted for MS Bishkek and Tyuya-Ashu, for MS Baitik the growth rate was slightly lower at 0.03 °C per year.

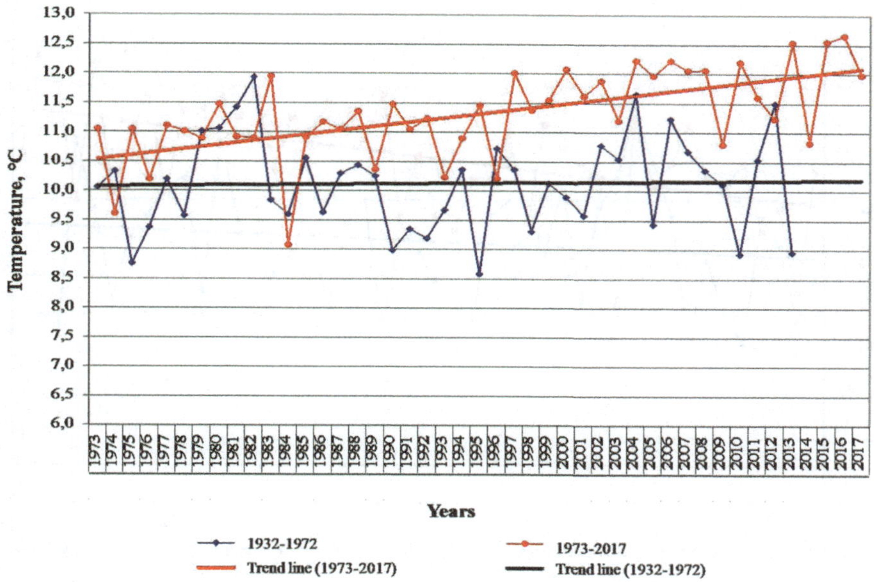

Fig. 5.2 Temperature trends according to Bishkek MS for the compared periods

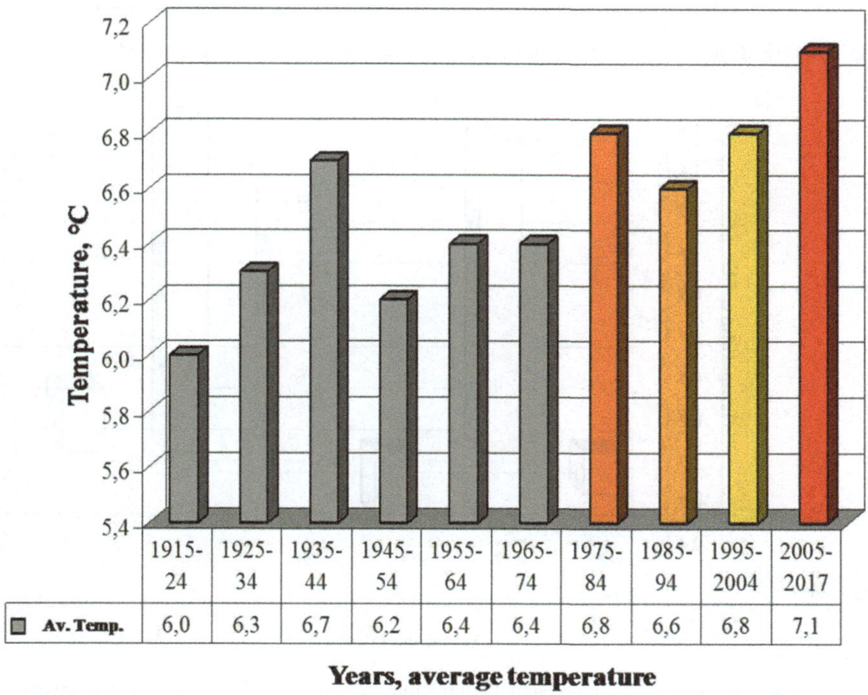

Fig. 5.3 Average temperature by Baitik MS over 10-year periods

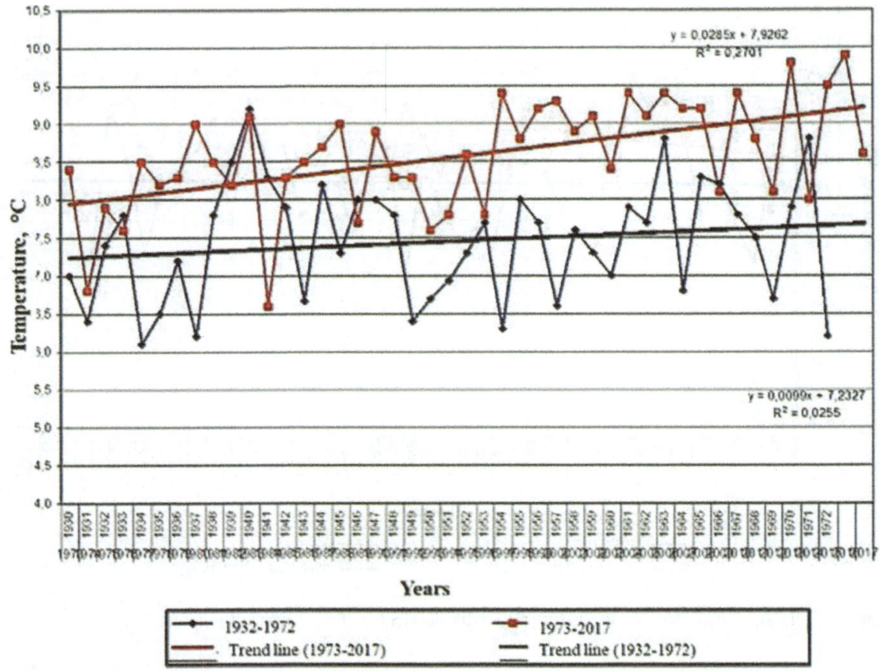

Fig. 5.4 Trends in average annual temperature according to MS Talas for compared periods

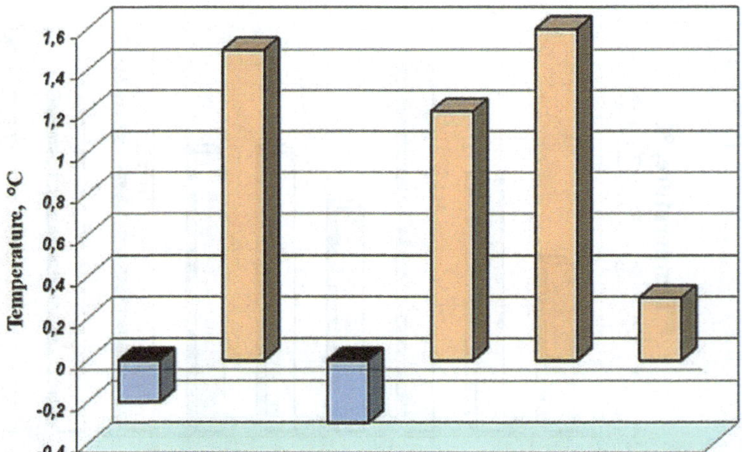

	2012	2013	2014	2015	2016	2017
Deviation	-0,2	1,5	-0,3	1,2	1,6	0,3

Years, deviations

Fig. 5.5 Deviations of the average annual temperature for 2012–2017. From the norm according to MS Talas

Table 5.1 Average annual temperature according to MS Talas for the observation period 1930–2017 and for comparable periods

Characteristics	Time period			
	1930–2017	1930–1972	1973–2000	2001–2017
Number of years	88	43	71/28	88/71/17
The average temp. °C/%	8.0	7.5	8.3	9.0
The difference in temp. °C	–	−0.5	0.3/0.8	1.0/1.5/0.7
The difference in temp. %	–	−6.2	3.8/11	12/20/8.4
Growth rate T°C/year	–	−0.10	0.004/0.029	0.011/0.021/0.041

Atmospheric Condensation

A trend analysis of the annual precipitation carried out according to the observational data gathered at these weather stations, showed that the change in precipitation is ambiguous, with both increases and decreases. At the same time, they also increased most significantly, by 70–80 mm in the valley part and especially after 1972 (Table 5.2).

From the data displayed in the table, we can conclude that the change in precipitation is ambiguous and different in both magnitude and by period. In the period between 1973 and 2017, their value varies from −9.0 to +51 mm, a percentage difference of −2.0 to +12% with an average of 16 mm or 7.6%. The most significant increase in precipitation according to weather stations, with the exception of Baitik, occurred over the past 17 years (2001–2017), from 13 to 83 mm (4–20%) and averaged 36 mm (9.2%).

South Region

This region includes pilot basins of the small transboundary rivers of the Syr Darya basin of Ak-Suu, Isfana, Isfara, and Padysha-Ata.

Table 5.2 The amount of precipitation according to the MS of the Chui basin for different periods

Time period	Weather stations		
	Bishkek	Baitik	ChuyaAshu (south)
1932–1972	405	567	789
1973–2017	456	560	797
Difference mm/%	*51* *12.6*	*7.0* *1.2*	*8.0* *1.1*
2001–2017	488	566	854
Difference mm/%	*83* *20.5*	*−1.0* *−0.2*	*65* *8.2*
1932–2017	429	563	793
For all entire period, mm	24	−4.0	4.0

Air Temperature

To analyze regional climate change using temperature and precipitation, three MSs with the longest series of observations located in different altitude zones were taken as references:

- Osh—1948–2017, altitude 888 m, plain zone;
- Uzgen—1940–2017, altitude 1012 m, foothill;
- Sary-Tash—1934–2017, altitude 3155 m, alpine.

The trend analysis of the average annual temperature at these weather stations for the observation period is unambiguously positive (Figs. 5.6 and 5.7).

The temperature increase from the initial and final trend values amounted to:

- According to MS Sary-Tash from −3.6 to −1.40 °C (difference of 2.20 °C);
- According to MS Osh from 11.2 to 13.1 °C (difference of 1.9 °C);
- For MS Uzgen from 10.7 to 11.90 °C (difference of 1.20 °C).

The most significant growth was noted after 1972. Figure 5.8 shows the air temperature trends for Sary-Tash MS for the compared periods, before and after 1972. The initial and final trend values changed in the first period from −3.5 to −2.5 °C, at

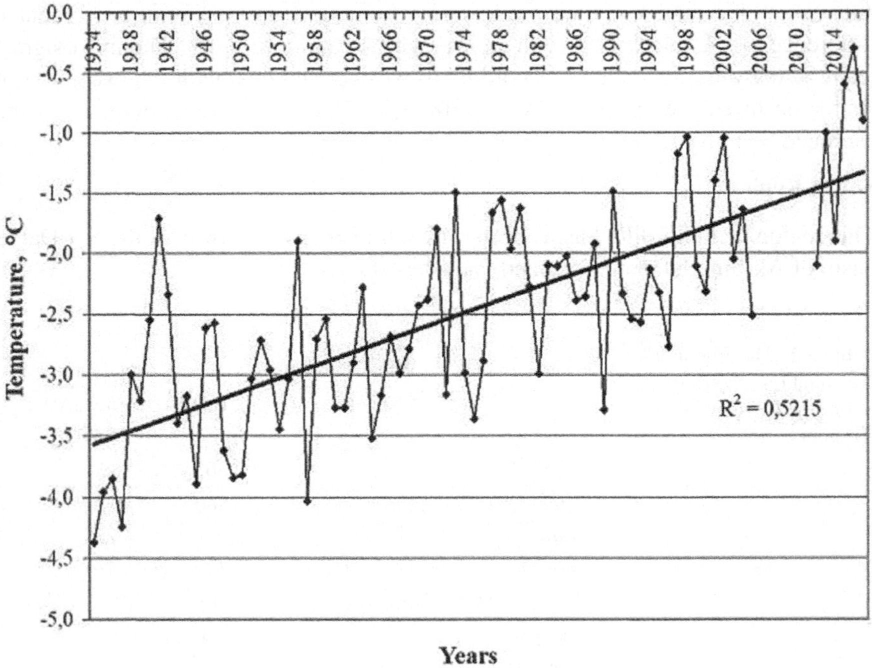

Fig. 5.6 Average annual temperature trend of (Sary-Tash MS)

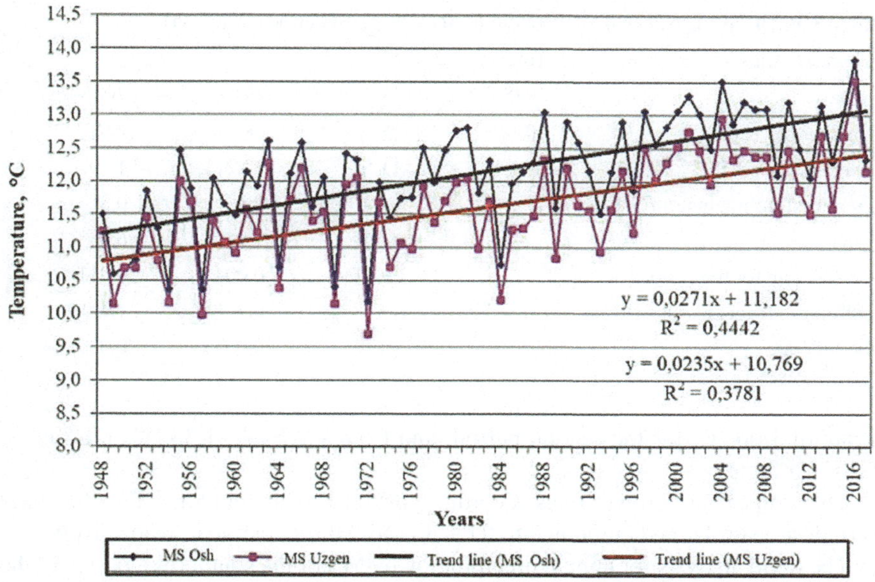

Fig. 5.7 Combined trends in average annual temperature (MS Osh and Uzgen)

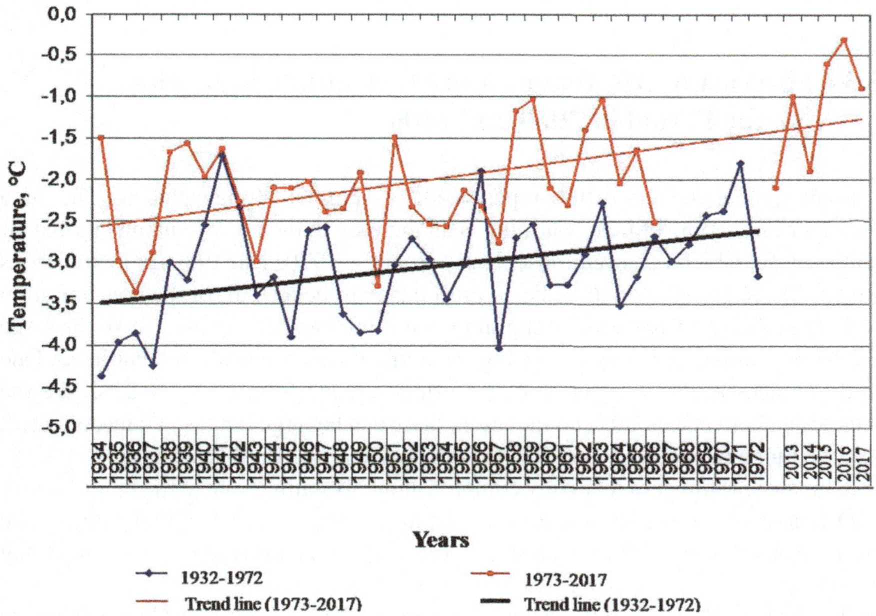

Fig. 5.8 Combined trends in average annual temperature. MS Sary-Tash for the compared periods

Table 5.3 The average annual temperature for the compared periods (MS Osh)

Characteristics	Time period			
	1948–2017	1948–1972	1973–2000	2001–2017
Number of years	70	25	28	44/17
Average period T°C	12.1	11.5	12.2	12.9
Deviation from the norm T°C	–	−0.6	0.1	0.8
Deviation %		−5.0	1.0	6.6
Growth rate for the period	–	−0.024	0.004	0.011
Difference from previous period	–	–	0.7	1.4/0.7
Growth rate		–	0.025	0.03/0.04

a rate of 1.00 °C, for the second period, and from −2.6 to −1.30 °C, at a rate of 1.30 °C (Table 5.3).

The temperature growth rates according to MS, Uzgen, and Sary-Tash for these periods are similar and amounted to 0.04 °C and 0.03 °C per year, respectively.

The trend analysis for annual precipitation by MS of the southern region over the observational period did not show a significant change, as in the Chui basin. There was a slight increase in the range of 6–9%.

5.3 Forecast of Air Temperature by Climate Scenarios for the Period of 2050 and 2100

Currently, there are very different perspectives on global climate change. The main indicator of ongoing climate warming is an increase in the average annual global air temperature, which occurred most clearly from the 1970s and 1980s [3]. Numerous studies of the global climate indicate this change to be a result of the concentration of CO_2 in the atmosphere doubling in the past 40 years. In order to show the large spread in estimates of global warming, modern climate models can be utilized. Due to differences in the description of important climate-forming feedbacks, like the interaction between clouds and radiation, there can be large ranges in data collected, like the range between 1.9 and 5.2 °C collected near the Earth's surface.

Climate warming is a relatively common anomaly dating back to the ninth century A.D. Scientists used observations of temperature changes at 14 different points of the Northern Hemisphere, the conclusion is clear. There is a period of global warming occurring.

The Kyrgyz Republic ratified the UN Framework Convention on Climate Change (2000) and its Kyoto Protocol (2003). Using global climate models, the basic scenarios of possible climatic changes in the Kyrgyz Republic for the period up to 2050 and 2100, it is very probable a result of an increased concentration of carbon dioxide in the atmosphere, as well as demographic and macroeconomic scenarios.

The result of international cooperation within the framework of this project was the 1st (2003) [5] and 2nd (2009) [6] national reports of the Kyrgyz Republic on climate change.

National reports of the Kyrgyz Republic on climate change. The 1s[t] National Communication presents trends in climatic characteristics like temperature and precipitation in Kyrgyzstan. It was created on the basis of long-term meteorological observations (until 2000). A forecast was given displaying the change of these characteristics with an increase in the concentration of greenhouse gases (1990 was taken as the base year). According to the results of the analysis for the entire territory of Kyrgyzstan, the average annual temperature in the twentieth century increased by 1.60 °C (in intervals of 0.6–2.4 °C), which is significantly higher than global warming of 0.60 °C. The greatest warming was observed in winter (2.60 °C) and the smallest in summer (1.20 °C). The increases were not the same for climatic regions and high-altitude zones. According to the estimate obtained for the entire period of instrumental observations from 1883 to 2005, the average temperature trend throughout the whole territory of the Kyrgyz Republic is 0.78 °C per 100 years. Precipitation throughout the Republic in the twentieth century increased slightly, by 23 mm or 6% [5].

In the 2nd report [6] forecasting climate change, climatic scenarios were created using the MAGICC/SCENGEN software package. As a result of preliminary analysis, versions of scenarios from the A2 and B2 families were adopted, which are characterized by more moderate socioeconomic indicators. Scenario A2-ASF gave the maximum value of CO_2 concentration in 2100. And in scenario B2-MESSAGE, the minimum value of CO_2 concentration was given by 2100.

The results of the calculated the temperature change (A2-ASF emission scenario) and (B2-MESSAGE emission scenario) are presented in Fig. 5.9. For each calculation region and two adopted scenarios, the dependence of the temperature change over the years are constructed with the allocation of minimum, maximum, and average model temperature changing. The names of the GCM (Global Climate Models) are given to determine the minimum and maximum temperature changes in 2020, 2050, and 2100 (Table 5.4).

In general, we can assume that in any case, a significant increase in average annual temperatures is expected. The A2-ASF emission scenario defines a larger increase in temperature change compared to the B2-MESSAGE scenario. For 2100, this difference is approximately 1.60 °C [6].

The 3rd National Communication from the UNDP Framework Convention on Climate Change already provides slightly different forecasts, not specifically for the territory of Kyrgyzstan, but global forecasts compiled using 4 climate models of the MSSP [7]. The periods for which the forecast is given are also different, 2046–2065 and 2081–2100. The forecast is based on the IPCC assessment report for a set of scenarios, namely, representative trajectories of RTK or RCP concentrations. According to the forecast, the average annual temperature for the first period will increase from 1.0 to 2.0 °C, with an interval of 0.4–2.6 °C, for the second period from 1.0 to 3.7 °C, with an interval of 0.3–4.80 °C.

In all three climatic scenarios, a significant increase in precipitation is not predicted.

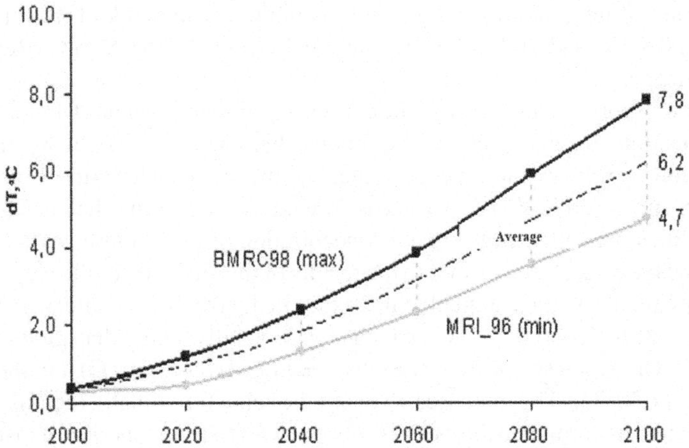

A2-ASF emission scenario

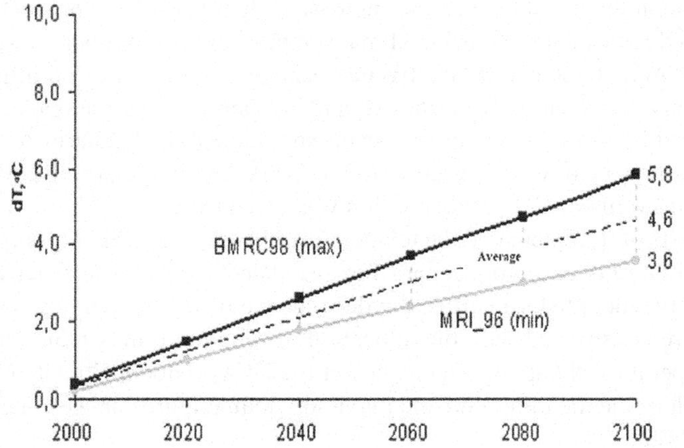

Emission scenario B2-MESSAGE.

Fig. 5.9 Changes (increase) in average annual temperatures relative to the base period (1961–1990)

Table 5.4 Northern region—Chui, Talas basins [6]

Emission scenario temperature rise A2-ASF		Temperature rise by scenario by scenario of emissions B2-MESSAGE	
2020	0. 3–1.2	2020	1.0–1.7
2050	1.8–2.8	2050	1.7–2.5
2100	4.7–7.8	2100	3.6–5.8

5.4 Temperature Forecast for 2050 and 2100 According to the Retrospective Analysis of Observation Data

Forecasting is a series of predictions with a certain lead time and the degree of accuracy concerning a particular climate element, in this case, air temperature. The development of forecasting methods and their further use allows one to determine the development of processes and phenomena in advance. Forecasts are based on observations of the previous period. This forecast method for such a long period (2050, 2100) can be taken as a consultation drawn up and based on a retrospective analysis of the data of the average annual temperature of the observational period, subject to the continuing trend and rate of climate warming over the forecast period [8].

In the approach used to forecast for such a long period of 34 and 84 years, the method of retrospective analysis of the data of the observation period, the inertia of the temperature change and the forecast of the rate of change during the forecast period were used.

In the process of analyzing the average annual temperature at all weather stations, it was found that the rate of temperature increase after 2000 increased, when forecasting for 2050 and 2100. For all weather stations, the period 2001–2017 was taken and the average temperature increase was 0.04 °C per year; the base temperature was taken as the average annual temperature for the period until 1973, the temperature increase until 2050 will be 1.40 °C, and until 2100 will be 3.40 °C. The values of the average annual temperature for the analyzed MS are presented in Table 5.5.

The forecast presented in the article is based on the data of instrumental observations for the entire period, according to which, using the trend analysis, the warming rate was calculated for the period after the 1970s to present and their extrapolation of the forecast period was carried out (2050, 2100). The findings of a trend analysis of average annual temperature and temperature growth rate resulted in the contention that warming will not be so significant compared to what previous forecasts predicted for climatic scenarios.

Table 5.5 Forecast of the average annual temperature for 2050 and 2100

MS	Average annual temperature (°C)		
	According to observations	Forecast	
	Before 1973	2050	2100
Bishkek	10.2	11.6	13.6
Baitik	6.4	7.8	9.8
Talas	7.5	8.9	10.9
TuyaAshuu	−3.9	−2.5	−0.5
Osh	11.5	12.9	14.9
Uzgen	11.2	12.6	14.6
SaryTash	−3.1	−1.7	0.3

Provided that the trend and pace (speed) of warming are maintained during the extrapolation of the trend of average annual temperature to 2050 and 2100, the predicted temperature will be increase on average 1.40 and 3.60 °C, respectively.

Climate change is especially significant in regional warming. The temperature increase is confirmed by the observational data of all supporting MS of Kyrgyzstan. Climate warming affects many ecosystems, whose condition depends on the temperature regime, like rivers, lakes, glaciers, soil moisture, biodiversity, etc.

References

1. Materials of the seminar and the international conference on climate change. Hydrol Ecol. Part 3.-Kaz NIIMOSK, 1997 (in Russian)
2. Budyko MN, Vinnikov KY, Drozdov OA (1978) Upcoming climate change. In: Proceedings of the USSR academy of sciences. Series geography, issue 6, pp 5–20 (in Russian)
3. Budyko MN (1978) Anthropogenic climate change. L. GIMIZ, p 405 (in Russian)
4. Scientific-applied reference on climate. Series 3. Perennial data. Parts 1–6, issue 32. Kyrgyz SSR, p 375 (in Russian)
5. The First National Communication of the Kyrgyz Republic on the UN framework convention on climate change. Bishkek, 2003, p 98 (in Russian)
6. The Second National Communication of the Kyrgyz Republic on the UN framework convention on climate change. Bishkek, 2009, 213 s (in Russian)
7. The Third National Communication of the Kyrgyz Republic on the UN framework convention on climate change is a draft. Bishkek, 2016, 252 s (in Russian)
8. Kuzmichenok VA (2003) Mathematical and cartographic modeling of possible changes in water resources and glaciation of Kyrgyzstan with climate change. News KRSU. T.3. No. 6, 2003, S. 53–64 (in Russian)

Chapter 6
Reactions and Dynamics of Drains on Small Transboundary Rivers with Various Feeding Types and the Effect on Climate Change

D. T. Chontoev and L. V. Bazhanova

Abstract The article presents the calculation results and analyzes the dynamics of water content in the rivers of small transboundary basins which are included in the CAREC project, "WATER, EDUCATION AND SCIENCE": the northern region—Kurkure-Suu (Talassky) and Aspara (Chui basins); southern region—Isfara, Ak-Suu and Isfana, Padysha-Ata (Syrdarya basin). Based on the results of a trend analysis and the extrapolation for the forecast period 2050–2100. The forecast-consultation of the average annual water discharge from these rivers is given.

Keywords Runoff · Transboundary-water · Linear correlation · Water resources management · Kurgyzstan

6.1 Introduction

Since the 1970s, climate change has been a topic of scientific discussions and has become a real problem, already requiring practical implementations in the modern period, forecasts for the future, and the development of measures to adapt to changing environmental conditions [1]. In Central Asia, climate change will affect river water availability, most of which (with the exception of the inland Issyk-Kul basin) are classified as transboundary rivers. A decrease in river water content due to the degradation of glaciers (the loss of glacial runoff) will lead to a deficit of surface water, which is vital in irrigated agriculture practices in the Central Asian Republics. This will lead to an aggravation of the problem of providing water to the populace and

---

Using the Examples of the Aspara and Kurkure-Suu Rivers.

D. T. Chontoev (✉)
Institute of Water Problems and Hydropower, National Academy of Sciences of the Kyrgyz Republic, 533Frunze St., Bishkek, Kyrgyz Republic
e-mail: doku66@mail.ru

L. V. Bazhanova
Laboratory of Water and Water-Energy Resources, Institute of Water Problems and Hydropower, National Academy of Sciences of the Kyrgyz Republic, 533 Frunze St., Bishkek, Kyrgyz Republic
e-mail: larisavas@mail.ru

Z. E. Kulenbekov and B. D. Asanov (eds.), *Water Resource Management in Central Asia and Afghanistan*, Springer Water, https://doi.org/10.1007/978-3-030-68337-5_6

economy of all states. As a consequence, there is a raised possibility of interstate conflicts.

This article includes research about a small basin of transboundary rivers the flow of which is used by two or three Central Asian states. It is there where cooperation is very important, this basin requires a joint approach to management, ensuring the rational use and adaptation of water resources to guarantee mutually beneficial conditions. Based on the results of a trend analysis for rivers is given for the forecast period 2050–2100.

The article uses hydrological and meteorological data from observational data of Kyrgyzhydromet [2], as well as published sources (monographs, articles, reports) on this topic.

6.2 Analysis of Water, Hydrological Justification, and Forecast of River Rivers for the Period 2050 and 2100

The river basin of the Aspara River forms a flow on the northern slope of the Kyrgyz mountain range along its northwestern border. The Aspara River is a tributary of the Kuragaty River, which then flows into the Chu River. The Aspara River basin is administratively located within the territory of two countries, Kyrgyzstan and Kazakhstan. This makes the basin a transboundary watercourse. The total catchment area of the basin is 1318 km^2, of which 876 km^2 are in Kyrgyzstan and 442 km^2 are in Kazakhstan.

According to the classification of V. L. Shchults [3] and the indicator of the type of feeding, δ is 1.56 (the δ-ratio is the runoff volume for July–September and the runoff volume for March–June) of the Aspara River refers to glacial-snow feeding. There is practically no glaciation in the river basin (0%), but the indicator of the type of feeding ($\delta > 1.0$) indicates the predominance of the summer runoff of the high mountain zone (more than 3000–3500 m a.s.l), is formed due to the melting of glaciers and snowfields. Significant areas in the river basin are occupied by buried glaciers (covered with moraine deposits) and snowfields (snow reserves that have not passed into the ice stage), which form the glacial component of the runoff.

For a trend analysis of the dynamics of river flow against the background of ongoing and predicted climate changes (temperature, precipitation, evaporation), a continuous long series of observations is required. With the Aspara River, the hydrological post of Kyryghydromet operated for 49 years (1927–1975) [4]. The selection of a river with a long series of observations (after 1975) to restore the interrupted series of observations did not yield positive results. For this reason, the trend of the average annual water discharge of the Aspara River was created only for the observation period and has a negative trend (Fig. 6.1).

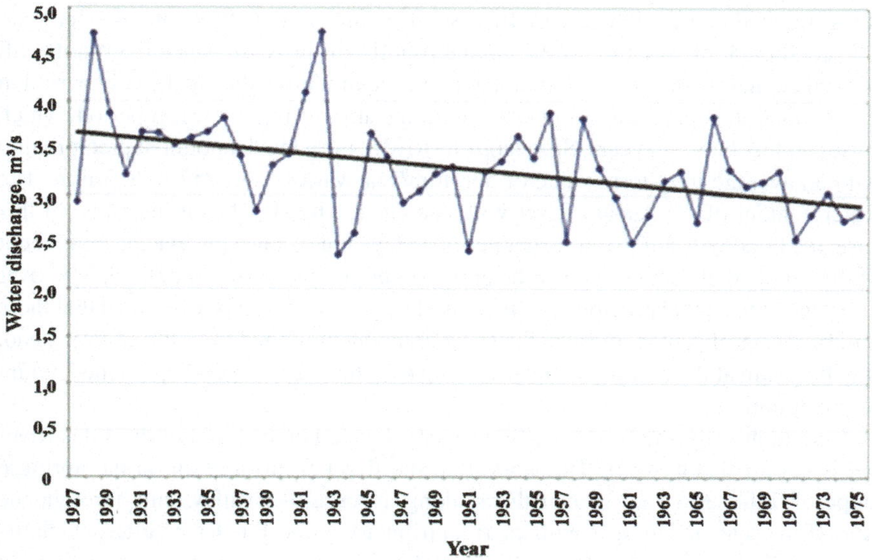

Fig. 6.1 The trend of average annual water discharge p. Aspara for the observation period

From the initial (3.6 m^3/s) to the final (2.9 m^3/s) trend value, the average annual water consumption decreased by −0.7 m^3/s (19%) over 49 years. The rate (intensity) of reduction was 0.014 m^3/s per year.

Based on the trend and the rate of decrease in runoff, slight glaciation will participate in the formation of runoff, but will not significantly affect the increase in water content, forecast, and consultation for the period between 2050 and 2100.

Table 6.1 shows the forecast values of the average annual flow rate for the period after the closure of the gauging station (1975–2000), as well as between 2050 and 2100. The average annual water flow rate of 3.28 m^3/s from the observation period of 1927–1975 was taken as the base.

According to the forecast, the average annual water discharge by 2050 will be 2.24 m^3/s, and by 2100—1.54 m^3/s. But like any long-term forecast, it is not subject to a quantitative assessment of justification and effectiveness.

During the observation period, the average annual flow rate of a dry year was 2.37 m^3/s, so the predicted flow rate by 2050 will be within this value, and by 2100,

Table 6.1 Forecast and consultation of the average annual runoff of the Aspara River

Period	Number years	Runoff reduction (m^3/s)	Water consumption according to the forecast (m^3/s)	Decrease from base expense (%)
1975–2000	24	0.34	2.94	−10
1975–2050	74	1.04	2.24	−32
1975–2100	124	1.74	1.54	−53

it is estimated that it will decrease by 0.83 m^3/s (35%) of the flow rate of a dry year. The coefficient of variation of the annual runoff will increase, since the runoff will be formed mainly due to precipitation (snow) accumulating during the cold period in the mountainous part and liquid precipitation (rain) during the warm period, which varies widely between years. So, according to MS, the annual amount of precipitation with an average long-term value of 538 mm/year varies from 257 to 855 mm. The water content of the Aspara River will also change by significant intervals. In this case, the flood will shift to an earlier period (May–June), since the glacial component of the runoff will decrease. Over the entire observational period for Baitik MS, over 100 years, the annual precipitation increased by 7%. So, it can be estimated that there will be no significant increase in the snow component of the forecast runoff, instead, it will remain at the current volume, varying over the years depending on the amount of precipitation.

Geographically, the Kurkureu-Suu River is located in the basin of the Talas River but is not a tributary of it. The Kurkureu-Suu River forms a drain on the northern slope of Talas Ala-Too, then it flows along a gorge-like valley, cuts through the Kara-Too ridge, and merges with the right tributary of the Ters River, after which the river is called the Assa River. Not far from the city of Taraz, the Assa River, without flowing into the Talas River, turns west and flows into the large Lake Bilyukol. Administratively, the river basin's location is divided between the territories of two states, the mountainous parts of Kyrgyzstan and the plains regions of Kazakhstan. Both countries use river flow for irrigation and water management needs, which classifies it as a transboundary river.

In the form of a hydrograph r. Kurkureu-Suu belongs to the Tien Shan type with two pronounced phases of the regime—high water (March–September) and low water (October–February). The value of the indicator of the type of feeding is 2.04, which indicates a significant glacial component of the runoff and classifies the river as one with an ice-snow feeding type [4]. According to the Catalog of Glaciers of the USSR (1970s), the river basin has a degree of glaciation of 3% (13.6 km^2 area).

A trend analysis of the average annual water discharge for the observational period (the total observational period is 80 years) showed that since the 1970s there has been an increased river flow, which continues into the present period of 2001–2017 (Fig. 6.2).

The initial value of the trend, collected in 1927, compared with the final value, collected in 2017, increased from 5.6 to 7.3 m^3/s by a rate of 1.7 m^3/s (30%). The average annual water discharges for different time periods calculated from the observational data also indicate an increase in runoff (Table 6.2).

The increase of water discharge, also called runoff, of the river is due to overall climate warming, as indicated by an analysis of the trends in average annual temperature according to the Talas MC for the observational period between 1930 and 2017 (Fig. 6.3).

The temperature increase between the initial and final trend values amounted to 2.1 °C, from 7.0 to 9.1 °C. The average temperature for the compared periods calculated from the observational data is presented in Table 6.3.

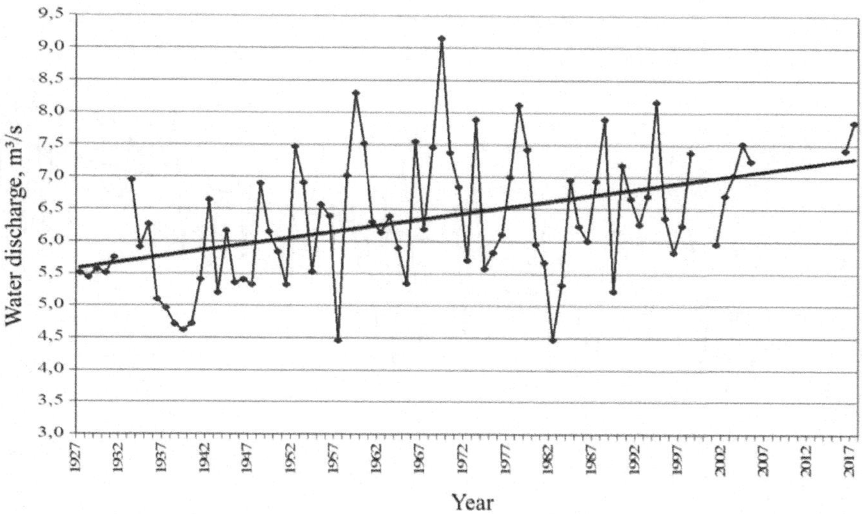

Fig. 6.2 The trend of average annual water consumption for the observation period

Table 6.2 Average annual water consumption for different periods

Characteristics	Time period			
	1927–2017	1927–1972	1973–2000	2001–2017
1	2	3	4	5
Average (m^3/s)	6.34	6.19	6.55	7.11
Difference (m^3/s)	–	−0.15	0.21/0.36	0.76/0.82/0.46
Difference (%)	–	−2.4	3.3/5.8	12/13/7.0

Note: In each billing period the following deviations are given:
Column 3: from the average value for the entire observation period
Columns 4 and 5: from the average values of previous periods

From the data provided in the table, it can be seen that the temperature increase is significant after 1972. For the period between 1973 and 2000, the temperature increased by 0.80 °C and compared with the previous period, the most significant increase was noted in the period after 2000.

From the results obtained, one can make an unambiguous conclusion, the trend of climate warming at this stage, after the year 2000, persists. Comparison between river flow and air temperature for individual periods is uniquely positive and the increase is noted (Table 6.4). It has been established that the average annual runoff of rivers with significant glaciation rises commensurate with an increase in air temperature.

According to the table, there is a linear correlation between the average annual temperature and water discharge for the compared periods (Fig. 6.4). The correlation coefficient reflects the tightness of the bond close to 1 ($K = 0.98$). This relationship confirms that the increase in the water content of rivers is due to the glacial-snow

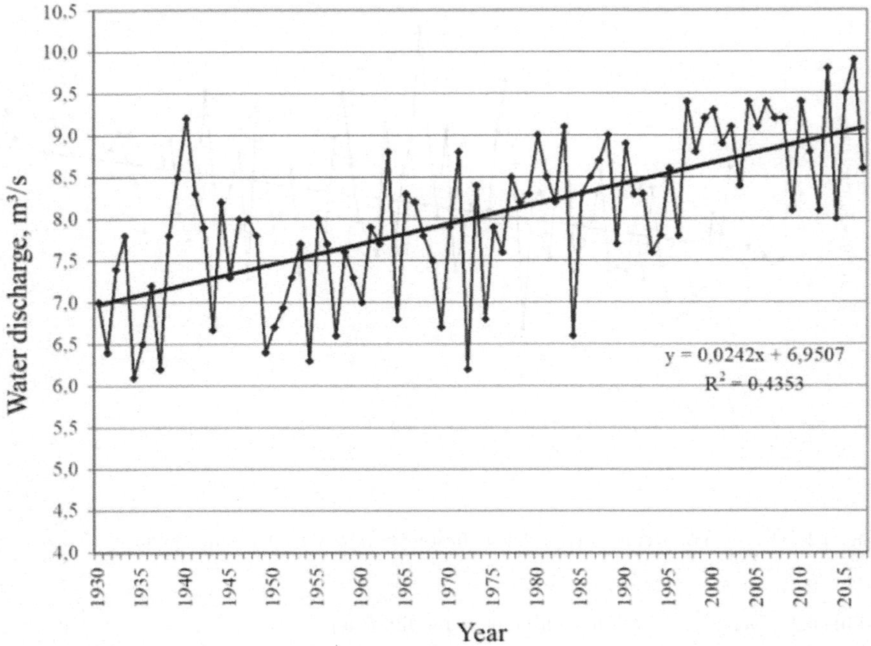

Fig. 6.3 The trend of the average annual temperature according to MS Talas for the observation period

Table 6.3 Average annual temperature for compared periods by MS Talas

Characteristics	Time period			
	1930–2017	1930–1972	1973–2000	2001–2017
Average	8.0	7.5	8.3	9.0
Difference (°C)	–	−0.5	0.3/0.8	1.0/1.5/0.7
Difference (%)	–	6.2	3,8/11	12/20/.4

Table 6.4 Comparative analysis of changes in average annual temperature (MS Talas) and water discharges of the Kurkureu-Suu river for comparable periods

Characteristics	Time periods			
	1930–2017	1930–1972	1973–2000	2001–2017
Average. temp. (°C/%)	8.0	7.5	8.3	9.0
Difference of temp (°C)	–	−0.5	0.3/0.8	1.0/1.5/0.7
Difference of temp (%)	–	6.2	3.8/11	12/20/8.4
Average discharge (m^3/s%)	6.34	6.19	6.55	7.11
Difference (m^3/s)	–	−0.15	0.21/0.36	0.76/0.82/0.46
Difference (%)	–	−2.4	3.3/5.8	12/13/7.0

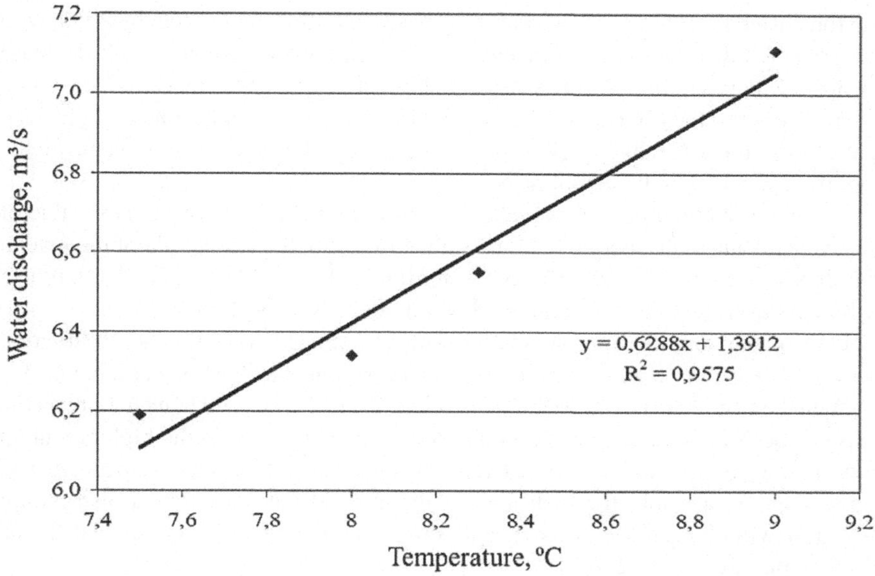

Fig. 6.4 The relationship between the average temperature of the MS (Talas) and water discharge (Kurkureu-Suu River)

supply and the more intense melting of said glaciers. The level of precipitation does not change significantly and, as a whole, does not have a decisive role in the formation of runoff.

The regression equation describing the relationship ($y = 0.629x + 1.39$) can be used to calculate the average annual runoff according to climatic scenarios for predicting air temperature for the period between 2050 and 2100. According to MS Talas, provided that the current growth rate of the average annual temperature of 0.030/year is maintained. However, by 2030, it will increase by 0.40 °C and amount to 9.00 °C. By 2050, it will increase by 1.20 °C, amounting to −9.80 °C. And lastly, by 2100, the growth will be 2.50 °C, making the average annual temperature 11.10 °C.

Using the temperature predicted in 2050, 9.80 °C, in the regression equation, the water discharge of the Kurkureu-Suu River should be 7.55 m^3/s. By 2100, with the predicted temperature of 11.10 °C, the flow should be 8.37 m^3/s. With such an increase in runoff, it is possible that with the participation of the glacial component the current volume will increase.

At the beginning of the 2000s, the possible consequences of climate change on the state of glaciers were studied and mapped at the Institute of Hydrometeorology and Geology (V. A. Kuzmichenok) [5, 6]. According to the calculations and maps, the number and area of glaciers in the Talas basin will decrease by 2050, and completely disappear by 2100.

When the hydrograph of the runoff of the Kurkureu-Suu River is divided into genetic components, snow, glacial, and underground, the glacial runoff (July–September) in the average water year is 53% of the total annual runoff [7, 8]. Utilizing the forecasted runoff rate 7.55 m^3/s for the base water flow rate. Then, by 2050, with the projected disappearance of glaciers, the average annual flow rate will decrease by 30%, which will be 2.27 m^3/s, and the flow rate will be 5.28 m^3/s.

According to the forecast, by 2100, with the complete disappearance of glaciers, water consumption will be reduced completely by the glacial component of the runoff, 53% which will be 4.44 m^3/s.

In another embodiment, if we take the observational period as the base value of the average annual discharge, 6.34 m^3/s, then by 2050, the magnitude of the glacial runoff will decrease by 1.90 m^3/s, the water flow will be 4.44 m^3/s. By 2100, runoff will be reduced to 3.36 m^3/s. and the flow rate to 2.98 m^3/s (Table 6.5).

For comparison, the average annual flow rate in a dry year for the observation period was 4.46 m^3/s, and the average annual flow rate was 6.34 m^3/s (Table 6.6).

With the complete disappearance of glaciers, the river flow will hang mainly on the seasonal accumulation of precipitation during the cold period in the high mountain zone (snow runoff) and associated rainfall during the flood period, from April–September. As a result, the hydrological regime and the intra-annual distribution of runoff will change, the flooding will shift to earlier months, and glacial floods (July–September) will decrease.

When developing recommendations on the adaptation in the field of water resources with the predicted reduction in terms of water use and water allocation, it is

Table 6.5 Forecast options for average annual runoff for 2050 and 2100

Average annual water consumption according to the equation of connection with temperature	Forecast for 2050		Forecast for 2100	
	Consumption reduction (m^3/s)	Average year consumption (m^3/s)	Consumption reduction (m^3/s)	Average year consumption (m^3/s)
7.55	2.77	5.28		
8.37			4.44	3.93
Average for observational period				
6.34	1.90	4.44	3.36	2.98

Table 6.6 Reduction in average annual water discharge of the Kurkureu-Suu river for the period 2050, 2100 relative to the average annual and average minimum values observation period

Forecast option	For the period 2050				For the period 2100			
	Relates; average year perennial		Relates; average year minimal		Relates; average year perennial		Relates; average year minimal	
	m^3/s	%	m^3/s	%	m^3/s	%	m^3/s	%
1	5.28	−17	4.46	18	3.93	−38	4.46	−12
2	4.44	−30	4.46	−1.0	2.98	−53	4.46	−33

necessary to focus on dry years that have already been observed over the observational period. According to the data of a trend analysis using both average annual temperature and temperature growth rate data, warming will not be particularly significant compared to the forecasts of climate scenarios. According to the forecasting results for all the most probable climate scenarios (2nd communication under the UNDP Framework Convention) [9], a significant decrease in surface runoff is expected. At the same time, an increase in surface runoff is expected in the period until 2020–2025. In the future, due to an increase in the impact of the glacial component, a decrease in runoff is predicted by approximately 43.6–88.4% of the runoff volume from 2000. In our opinion, the value (88.4%) is overstated.

6.3 Conclusion

According to the forecasting results, by 2050, a significant decrease in surface runoff is expected for transboundary rivers of small basins, regardless of the type of feeding. The reduction in average annual water consumption for all the rivers under consideration will be phased from 20 to 52%. The reduction will occur due to the reduction of the glacial component of runoff, subject to the partial and complete disappearance of glaciers by 2050 through until 2100.

Under climate scenarios (2nd UNDP Report on Climate Change in Kyrgyzstan), surface runoff is expected to increase until 2020–2025. Due to an increase in the glacial component, by the period between 2050 and 2100, it is expected to decrease to approximately 42.4–20.4 km^2, which is 43.6–88.4% of the runoff from 2000 on all rivers of the Kyrgyz Republic [9]. According to the forecast given in the article, the reduction in runoff will not be as significant as previously thought.

Meanwhile, the projected reduction in surface runoff, will certainly affect the living conditions and economic activity, both in Kyrgyzstan and in neighboring countries, as the rivers and their resources are transboundary. Reducing the glacial component of runoff will have a significant impact on the intra-annual distribution of river runoff, significantly reducing its summer maximum and shifting it to an earlier date. In the mountain zone, glaciers accumulate atmospheric precipitation in solid form year-round and they contribute most of the water in the important summer months for agricultural production. Glaciers also help to increase the river flow in the dry, hot years as well, helping to compensate for the low rainfall in arid years.

Glaciers as water resources in Kyrgyzstan are moisture accumulators that distribute runoff over the entire flood period. A reduction of water availability without adopting appropriate adaptation measures can significantly affect the main consumers of transboundary water resources of Kyrgyzstan and neighboring Central Asian states.

Climate change adaptation should focus on:

- Improvement of water resources management, implementation of the principles of integrated water resources management (IWRM);

- improvement of the legislative framework and institutional changes;
- maintaining the formation zone, both at the national level and in a transboundary context;
- rational use of water resources, which involves the development and implementation of economic incentives and the rehabilitation of water facilities;
- restoration and expansion of the monitoring system for natural and climatic factors;
- construction of new water facilities with a view to rational distribution of water throughout the growing season;
- raising awareness about the state of water resources, both of decision makers and the public;
- increasing the personnel potential, from monitoring observers to water managers to lawyers in the water law field.

References

1. Climate change and its consequences. St. Petersburg. “Science”, p 269 (In Russian)
2. Kyrgyzhydromet data, https://meteo.kg/login
3. Schulz VL (1965) The rivers of Central Asia. L.: Gidrometeoizdat, p 691 (In Russian)
4. Resources of surface waters of the USSR. T.14, issue 2. The basin of Lake Issyk-Kul, Chu, Talas, Tarim rivers. Hydrometeoizdat. L. 1973 (In Russian)
5. Kuzmichenok VA (2003) The volume of glaciers in Kyrgyzstan (assessment, changes, forecast). Materials of the seminar “ecology and clean water”. Soros Foundation, Issyk-Kul (In Russian)
6. Kuzmichenok VA (2003) Mathematical and cartographic modeling of possible changes in water resources and glaciation of Kyrgyzstan with climate change. Vestn. Kyrgyz-Russian Slavic University. No. 6, S. 53–64 (In Russian)
7. Goroshkov IF (1979) Hydrological calculations. L.: Godrometeoizdat, p 431 (In Russian)
8. Chebotarev AI (1964) Hydrological dictionary. L.: Gidrometeoizdat, p 222 (In Russian)
9. 1st and 2nd National Communication of the Kyrgyz Republic on the UNDP framework convention on climate change in Kyrgyzstan. Bishkek, 2003, 2009 (In Russian)

Chapter 7
The Glacial Systems of Kyrgyzstan Under Climate Change

Ryskul Abdyzhakypovich Usubaliev, **Bolot Duishenalievich Moldobekov**, **Alexander Nikolaevich Mandychev**, **Azamat Tynarbekovich Osmonov**, **Erlan Azisov**, **Ruslan Kenzhebaev**, **Muhambet Esenaman uulu**, **and Anton Gennadievich Shabunin**

Abstract The conditions, the main stages of development, and the evolution of glaciation of the mountain systems of Kyrgyzstan have been thoroughly analyzed. The tendency of changes in the basic climatic elements—air temperature and precipitation has been studied at two high-altitude meteorological stations of the Tien Shan and Gissaro-Alai. The results of the monitoring of the mass balance of representative glaciers has been presented and the current scales of degradation and retreat of the Kyrgyz glaciers under conditions of modern directed warming of the surface layer has been analyzed. The air temperature and precipitation trends have an asynchronous course, a negative sign dominates in the balance of glacier mass in the vast majority of the Tien Shan and Pamir-Alai mountain systems as a consequence of global climate warming. As a result, the degradation of glaciers intensifies, and they shrink with different intensities depending on their location in mountain systems. Based on satellite images of "Landsat-8" from 2013 to 2016, an inventory of glaciers has been taken, an updated Catalog of Kyrgyzstan glaciers have been compiled and published.

Keywords Glacier systems · Glacier · Tien Shan · Gissaro-Alai · Glacier catalog · Climate change · Mass balance · Degradation · Fluctuations and glacier shrinkage

7.1 Introduction

Kyrgyzstan is distinguished by its well-developed glaciation in Central Asia, being located inside the Eurasian continent, far from the oceans and in the zone of temperate latitudes. There is a big part of the western component of the Tien Shan, a huge mountain system, and parts of the northern and eastern periphery of the powerful Pamir-Alai mountain system within its territory. The territory of Kyrgyzstan is categorized as highlands. More than half of its territory (56.9%) is located at altitudes

R. A. Usubaliev (✉) · B. D. Moldobekov · A. N. Mandychev · A. T. Osmonov · E. Azisov · R. Kenzhebaev · M. Esenaman uulu · A. G. Shabunin
Central Asian Institute of Applied Geosciences, Timur Frunze Rd.73/2, Bishkek, Kyrgyzstan
e-mail: r.usubaliev@caiag.kg

Z. E. Kulenbekov and B. D. Asanov (eds.), *Water Resource Management in Central Asia and Afghanistan*, Springer Water, https://doi.org/10.1007/978-3-030-68337-5_7

over 2500 m and almost a quarter (23.0%) above 3500 m [1]. Almost all mountain ranges of Kyrgyzstan, to a greater or lesser extent, have centers of modern glaciation. The glacial systems of the mountain ranges of Kyrgyzstan are not only an element of the alpine landscape, but they are also significant water resources accumulating in the solid phase and reservoirs of fresh water. Glaciers are significant water resources accumulating in its solid form, acting as a reservoir of fresh water. This feature determines the regulatory role of glaciers in the process of river runoff formation. The snow and ice resources concentrated in them are released during the warm seasons. Their role has increased, particularly in years with insufficient moisture, and in dry years' glaciers become the only sources of nutrition for rivers, and thereby providing water resources to the lower valleys and plains.

7.2 Peculiarities of the Existence and Development of Glaciation in the Mountains of Kyrgyzstan

One of the leading factors in the formation and existence of modern glaciation for the Tien Shan is terrain. The connection [2, 3] between the height of the ridge and the possibility of existence, development, or reduction of glaciation under modern climatic conditions has been established. Glaciation can only be on ridges with elevations of more than 4000–4100 m. The larger the area of the ridge at altitudes above 4000 m, the greater the glaciation. There are no glaciers on ridges with a maximum height of up to 3700 m. Despite the large number of ridges, due to differences in their orographic structure and heights, the sizes of modern glacial systems can vary greatly. On some ridges, the area of glaciation is units of square kilometers, on the others, possibly even thousands of square kilometers. There are 5237 glaciers covering an area of 6336.1 km^2 or 63.4%, respectively, of the total number and 71.7% of the total area of the glaciation of the western component of the Tien Shan, which belongs to three countries in Central Asia, Kyrgyzstan, Kazakhstan, and Uzbekistan, within the Kyrgyz Tien Shan, according to the catalog of the USSR glaciers, compiled in the 1960s to the 1970s [3].

The Pamir-Alai is divided into two mountain systems, the Pamir and Gissaro-Alai, according to peculiarities of the geological structure, topography, and climate [4]. They are separated by a wide longitudinal depression occupied by the Vakhsh River Valley, which, in the upper reaches is called Kyzyl-Suu. The differences between the Pamir and Gissaro-Alai are noticeable. The leading factor is the existence and development of modern glaciation in the Pamir is terrain, while in the Gissar-Alai it is ridges, modern glaciation is more significant due to joint factors of climate and terrain. The northern slope of Turkestan (the eastern, highest part of the ridge), the northern and southern slopes of the Alai ridge of Gissaro-Alai, as well as the northern slope of the Chon-Alai Ridge of the Pamir are all located within the territory of Kyrgyzstan. The last ridge is 1000 m higher than the Alai Ridge. The entire northern slope of the Chon-Alai ridge, with the exception of minor spurs in the extreme west and east, is located in Kyrgyzstan and is the southern border of the Alai Valley. There are 1327 glaciers with an area of more than 0.1 km^2, their total area is 1684.5 km^2 in the Kyrgyz part of the Pamir-Alai according to the Catalog of Glaciers of the USSR.

There were 7628 glaciers covering an area of 8107.2 km^2 within the territory of Kyrgyzstan, based on the data from the USSR Glacier Catalog [3]. This is 4% of the overall territory of Kyrgyzstan. The volume of ice accumulated in glaciers, only on the Tien Shan ridges, was estimated to be 620 km^3 [5].

Mountain glaciers, as a natural object in the form of ice and snow, are formed as a result of the favorable combination of terrain conditions and climate. In the history of the evolution of glaciation of mountain systems, there were noted repeated periods of activation and degradation, (i.e., during climate warming or cooling, glaciers changed their size and volume). Consequently, the scale of glaciation also changed. Various research studying glaciers in the mountains of Central Asia is a relatively recent development. From a geological point of view, the scale of glaciation changed from 2 to 5 ice ages. For example, there has only been three cases of glacier activation over the past 60,000 years, in the periods of 60–50, 35–25, and 20–14 thousand years ago. At the maximum development of glaciation, glaciers descended down into the modern valleys. At the same time, the total area of glaciation exceeded the modern area by 3–4 times [6]. After repeated Pleistocene intensifications of glaciation, which were distinguished by enormous proportions, in the Holocene, the glaciation collapse begins in the period of 10–12 to 12–15 thousand years ago. The climate of the Holocene differs little from the more ancient interglacial times; therefore, it represents a typical interglacial epoch. The trend of climate change in the Holocene begins with the transition from the cold conditions of the end of the Pleistocene to a warm climatic optimum, and then to a new cold spell. Two warm periods were noted in the subsequent climatic optimum, where the maximum of warming falls 6000 years ago. Lowering the temperatures after each warm period led to the activation of glaciers. But, the scale of the movement of the glaciers was not as significant as before. The next cool and humid sub-Atlantic period (2200–2500 years ago) was marked by the expansion of the permafrost zone, and its end was between the fifteenth and nineteenth centuries A.D. During this period, there was an increase in glaciation in many mountainous and polar regions. It is precisely on the interval of these centuries the last noticeable advance of glaciers falls and, in connection with this, this period of time is sometimes called the "Little Ice Age", or "neoglacial" time [7, 8]. During this period, there is a surge in the activation of glaciers in the second half of the seventeenth and eighteenth centuries, as well as at the beginning, middle, and end of the nineteenth century. After the end of the last "little ice age" (1450–1850 A.D.), the glaciation of the Tien Shan and Pamir-Alai passes into the next stage of degradation with periods of short-term activation of the glaciers.

7.3 Glacier Dynamics Under Conditions of Directed Climate Warming

Modern climate warming in Central Asia began to appear in the mid-1960s [9]. The ongoing global warming of the surface layer and the ambiguous change in precipitation in the alpine zone, they either decrease or remain within the climatic norm, are manifested in a steady reduction in modern glaciation. Despite regional differences in the climatic conditions of Kyrgyzstan, the background climate is the same for the entire Central Asian region and there is a general trend of glaciation degradation, covering the entire territory of its mountain systems.

A change in the regime of the main climatic elements (air temperature and precipitation) determines one of the reasons for the change in the spatial position of glaciers. One of the main tasks while researching the instability and stability of glacial systems is to study the regime of air temperature and precipitation under conditions of a continuing trend in climate warming.

Of particularly great practical and scientific interest is the nature of temporary instability and the current trend in the dynamics of temperatures and precipitation. There are only two meteorological stations (MS) with relatively long observational series (70–90 years) currently operating in the alpine zone, MS Tien Shan (N = 3614–3659 m), and Sary-Tash (N = 3155 m.). The first one is located in the upper Naryn region in the Inner Tien Shan. The second is located in the eastern part of the Alai ridge.

There was a general trend of an increased average annual temperature by 2.4 °C on the MS Tien Shan within the 85 years between 1930 and 2015. The range of fluctuations in average annual temperatures was 6.5 °C and the air temperature in the last decade was mainly above the norm (see Fig. 7.1). On the Alai Ridge, according to MS Sary-Tash, for the period from 1934 to 2011, the temperature of the surface layer in the alpine zone increased by 1.95 °C. This is relatively more than in the mid-mountain areas of the Inner Tien Shan. So, the long-term average annual air temperature increased by 1.05 °C in the Middle Naryn Valley of the Inner Tien Shan over the past 105 years (1894–1999) [10].

Precipitation is one of the most unstable elements of the climate. Long-term fluctuations in precipitation are connected mostly with the instability of synoptic processes in a given year. Despite the fact that synoptic processes of the same type cover a large territory, the intensity of atmospheric precipitation changes in the Tien Shan and Pamir-Alai, due to the features of the terrain structure, orography and hypsometry of the ridges.

Analysis of the chronological course of precipitation at MS Tien Shan shows that the amount of precipitation is subject to fluctuations. Rarely, their values are equal to the average long-term sum (norm), mostly precipitation was less or more than the norm. The downward trend in precipitation is noted. (Fig. 7.2).

For 85 years, at MS Tien Shan, the trend decrease in the amount of precipitation during the warm period comprised of 20.3 mm. With the norm of atmospheric precipitation of the warm period being 252.7 mm, the decrease in precipitation was

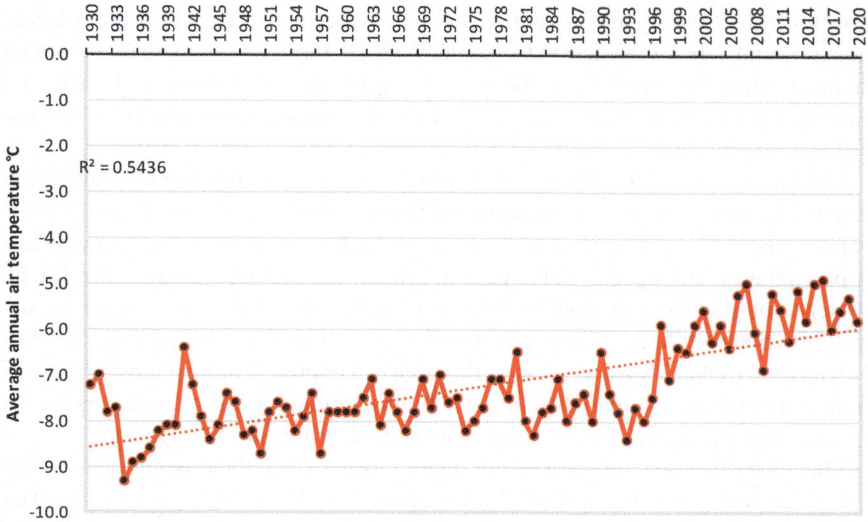

Fig. 7.1 The time course of the average annual air temperature at the MS "Tien Shan"

8%. However, since 1998 there has been a sharp increase in the amount of precipitation. This is due to the transfer of the weather station to another location and to a comparatively higher altitude. But, despite the increase in the amount of precipitation over 18 years, the trend decreases in the amount of precipitation in the warm period, amounting to 41.7 mm, i.e., precipitation decreased by 14.3% with a normal precipitation being 292.4 mm. With precipitation at MS Sary-Tash, a different picture is

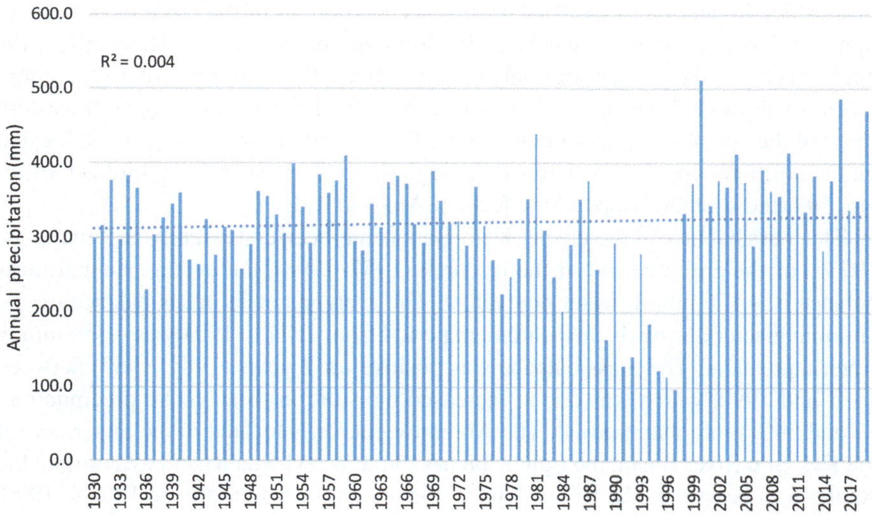

Fig. 7.2 The time course of the average annual precipitation at MS Tien Shan

observed. The high mountain zone of the Alai Ridge is characterized by increased moisture, and here for the period from 1934 to 2010, there was an increase in average annual precipitation by 64.3 mm. While in the highland zone of the Inner Tien Shan, it was the opposite, there was a decrease in long-term annual precipitation by 19.4 mm [10].

For the area of MS Tien Shan, with a spring–summer maximum of humidification, an asynchronous course of temperatures and precipitation is characteristic. That is, warming is accompanied by a decrease in precipitation, mainly due to a decrease in precipitation during the warm period, which is extremely unfavorable for the existence of glaciation in modern climate warming. As for the Sary-Tash area, it is again the opposite, there is a synchronous course of temperature and precipitation with upward trends. However, despite this, due to a relatively more intensive increase in air temperature, compared with other high mountain areas, the glaciers on the Gissaro-Alai Ridges are greatly degraded, as in other parts of Kyrgyzstan.

The most objective indicator of the state of glaciers and their evolution are long-term data on the mass balance of glaciers. This value has been negative in the Tien Shan and Pamir-Alai glaciers since the beginning of the 1970s, with the exception of the rare positive mass balance due to individual years connected with favorable weather and climatic conditions. Long and continuous annual measurements of the mass balance of glaciers in Soviet times were carried out on only three glaciers in Kyrgyzstan. These are the Kara-Batkak glacier (the northern slope of the Teskey Ala-Too range), with its mass balance being characterized as the inner Tien Shan, in particular the Issyk-Kul Basin. The Golubin glacier (the basin of the Ala-Archa River, on the northern slope of the Kyrgyz Ala-Too range) characterized by the mass balance of the glaciers of the Northern Tien Shan. And lastly, the Abramov glacier characterized by the mass balance of the Gissaro-Alai glaciers (on the southern slope of the Alai Range). However, by the end of the twentieth century, observations and other glaciological types of work were suspended on all of the three of the above-mentioned glaciers due to financial difficulties and other reasons. The work on the mass balance of the glaciers was only restored after 2010. Moreover, two–three more glaciers were added. The first one is Glacier No. 354, it is located in the northwestern slope of the Ak-Shyyrak mountain massif. The second glacier is West Suek, located on the northern slope of the Jetim-Bel Ridge (Inner Tien Shan), the third is Glacier No. 599 (the southern slope of the Kungei Ala-Too Ridge).

The mass balance of the glacier Kara-Batkak has been consistently negative since 1973. For the entire period of the research (1957–1998), the average annual mass balance is −436 mm in water equivalent. The average annual mass balance on the Golubin glacier during the measurement period from 1971 to 1991 was −359 mm in water equivalent. The mass balance was positive for 2 years (1981, 1987) between 1981 and 1990, which was due to increased moisture in these years: precipitation, 37 and 59% more than normal. The Abramov glacier is located in the upper part of the Kek-Suu River Basin, the right tributary of the Kyzyl-Suu River (western) on the southern slope of the Alai Ridge. During the observation period from 1968 to 1990, the average annual mass balance was −70 g/cm^2. A positive glacier mass balance was observed only in 1969, 1972, and 1987 [10].

Table 7.1 Mass balance of representative glaciers of Kyrgyzstan, (mm.w.e.)

Glaciers	2010–2011	2011–2012	2012–2013	2013–2014	2014–2015	2015–2016	2016–2017
Abramov	–	−601	−249	−665	−33	+380	−66
Western Suek	−371	−470	−363	−459	−822	−425	−872
No. 354	−356	−463	−463	−675	−635	−399	−574
Golubin	−48	−143	113	−1669	−489	+287	−144
No. 599	–	–	–	–	−62	−725	−197

It is turning now to the latest results. Table 7.1 shows the mass balance of the studied glaciers in the corresponding hydrological years. As follows from the table, the balance is negative, that is, ice consumption exceeds the amount of accumulation. But, there are exceptions. In the 2015/16 balance year, the mass balance of the Abramov and Golubin glaciers was positive.

Due to the frequent excess of side expenditures, the balance of the incoming, both the thinning of the surface and retreat of the glacier front processes are observed directionally. In general, for all the studied glaciers, retreat of the boundaries of the tongues continues and accordingly, there is a decrease in the volume of ice and a decrease in the reserves of water accumulated in the glaciers. However, for individual glaciers, there is a temporary decrease in the speed of degradation in the form of a slight retreat of boundaries and a positive mass balance. The dynamics of the fluctuations of glaciers and the trend in their advancing have been studied by many researchers [11–18] and others]. According to the results of numerous studies, the reduction in the area of the Tien Shan and Gissaro-Alai glaciers from the mid-nineteenth century to the present day has been about an average value of $20 \pm 10\%$, which, taking into account the observation period of about 150 years, gives a rate of change of about 0.07–0.2%/year. In particular, this is evidenced by observational data from the Abramov Glacier from 1850 to the present day, its area has decreased by 13.8%. The average speed of reduction of the glacier area over the entire period under consideration is about 0.02 km^2/year. Over the same period of time, the length of the glacier decreased by approximately 2950 m. With an average speed of about 18 m/year [14], (Fig. 7.3).

Considering the changes in the glacial systems of the macroscopic slopes of individual mountain ridges and using the example of the northern slope of the Teskey Ala-Too Ridge, it is possible to track the evolution of glaciation over a short period of time under the conditions of current directed climate warming, defining the current area of glaciers. The decoding of satellite images of Planet Scope satellites with resolutions of 3.0 m per pixel occurred on August 2018. The results of the analysis of satellite imagery showed that, in 2018, on the northern slope of the Terskey Ala-Too Ridge, glaciers with an area of more than 0.1 km^2 were detected in the amount of 633 units with a total area of 384 km^2. Table 7.2 shows that there are more glaciers with the size of less than 0.5 km^2 on the northern slope of the Terskey Ala-Too Ridge, however, their area is much smaller than the area of larger glaciers.

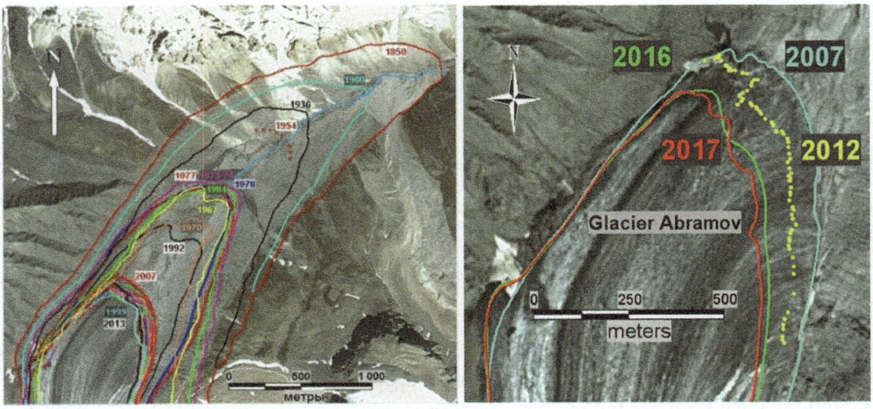

Fig. 7.3 The boundaries of the Abramov glacier in different years from 1850 to 2017

Table 7.2 Current figures and area of glaciers of the northern slope of Terskey Ala-Too

Glacier sizes	Quantity	in %	Area (km^2)	in %
<0.5	441	69.7%	82.2	21.4%
0.5–1.0	78	12.3%	56,9	14,8%
1.0–5.0	108	17.1%	203	52,8%
5.0–10.0	5	0.8%	31.8	8.3%
>10.0	1	0.2%	10.2	2.7%
	633	100	384	100

For comparison, we presented the results of work performed several years earlier on the basis of satellite images of "Landsat-8" in 2013 and data from the Catalog of Glaciers of the USSR. So, according to the results of the analysis of satellite images "Landsat-8" in 2013, on the northern slope of the Terskey-Ala-Too Ridge, there were 499 glaciers with size of more than 0.1 km^2, with a total area of 447.5 km^2 [19]. According to the Catalog of Glaciers of the USSR [20], there were 481 glaciers with size of more than 0.1 km^2, with a total area of 496.7 km^2. In total, there were 675 glaciers with a total area of 510.1 km^2, including 194 glaciers with the size of less than 0.1 km^2, with a total area of 13.4 km^2, in the glaciation area of the northern slope of the Terskey Ala-Too Ridge. Thus, the total area of glaciers is decreasing directionally, and the number of glaciers is growing due to the collapse of glaciers into various smaller sizes.

A similar picture is observed in Kyrgyzstan on other ridges and regions of the Tien Shan and Pamir-Alai. In 2018, an inventory of Kyrgyzstan's glaciers based on satellite imagery was completed [19]. In total, according to satellite images of "Landsat-8" as of 2013–2016, there are 9959 glaciers in Kyrgyzstan with a total area of 6683.9 km^2, including: 6227 glaciers with size of than 0.1 km^2, with a total area of 6494.0 km^2 and 3732 glaciers with size of less than 0.1 km^2, with a total area of 189.9 km^2.

According to the Catalog of Glaciers of the USSR (40–70 s of the XX century), in Kyrgyzstan there were 8164 glaciers with a total area of 7944.2 km^2, including: 6719 glaciers with size of more than 0.1 km^2, with a total area of 7866.6 km^2 and 1445 glaciers less than 0.1 km^2, with a total area of 77.6 km^2. The total number of glaciers increased by 22%. This is due to an increase in the number of small glaciers (with the size of less than 0.1 km^2) by approximately two and a half times (258%), while the number of large glaciers (with the size of more than 0.1 km^2) decreased by 7.5%. Information on the reduction of glaciers and current distribution in the main river basins of Kyrgyzstan is shown in Figs. 7.4 and 7.5.

The reduction in the area of the main part of the glaciers of Kyrgyzstan is in the main basins is within 13–17%. The exception is the Talas River Basin, where the reduction in area reaches 47%; the Chu River Basin reaches 28%; the Amu-Darya River Basin reaches 10%; in the Karadarya River Basin, glaciation has not changed; and the Chatyr-Kul Lake Basin, since glaciation in this basin is practically stable (there are 9 glaciers in total with a total area of 3.2 km^2). As can be seen from the diagram, almost half of the glaciation is on the Tarim River Basin (45%) and the third part is on the Syr-Darya River Basin (30%).

Thus, over the approximately 70-year period, the following changes have occurred in the general glaciation of Kyrgyzstan. The area of glaciation decreased by 16%, while the area of large glaciers decreased by 17%, while the area of small glaciers increased by two and a half times (245%). This is due to the general degradation of glaciation, in which the degradation of large glaciers leads not only to a decrease in

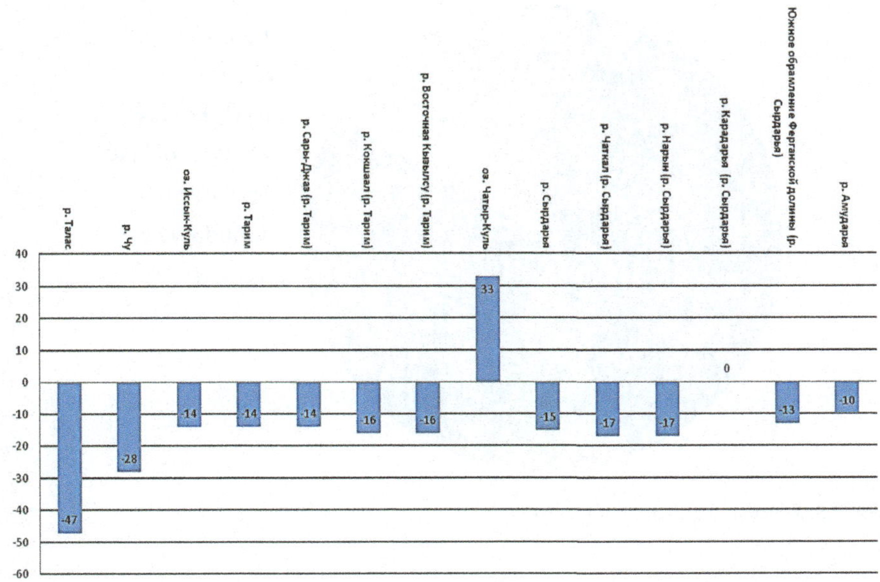

Fig. 7.4 Reduction in the area of glaciers in the main river basins of Kyrgyzstan from 40–70th of the twentieth century to the present (%)

their area, but also to their disintegration into separate parts that function as independent small glaciers. In addition, in some parts of the USSR Glacier Catalog, glaciers with sizes less than 0.1 km^2 are not taken into account at all, and in this catalog all the glaciers with a size of more than 0.01 km^2 each are taken into account.

7.4 Conclusion

Currently, an unfavorable combination of climatic conditions for the existence and advancing of glaciation in the Tien Shan and Pamir-Alai continues. Despite the fact that in the glacial-nival zone, precipitation is 2–3 times more than in the lower valleys and plains, an increase in air temperature is noticeably seen in glacier regime, especially in the discharge part. The mass balance of glaciers in most cases is negative. A rare positive mass balance of glaciers is associated with precipitation that has fallen above normal. Glacier degradation intensifies and retreats with different intensities, which leads to a significant reduction in the size of glaciers. The size of the glaciers is intensively shrinking not only in length and area, but also there is a thinning of the surface of glaciers. Moreover, not only the ablation zone, but also some parts of the accumulation zone are subject to reduction, since in many ridges the snow line has gone beyond the maximum elevations of the ridge, with the exception of some, the highest levels of the glacial basin.

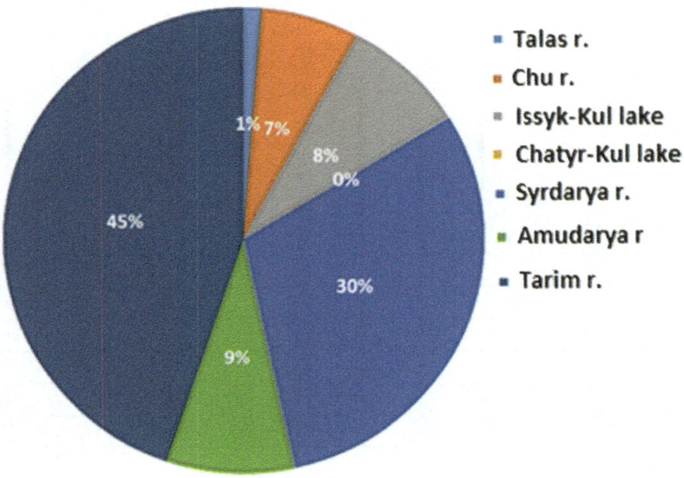

Fig. 7.5 Current distribution of glacier areas

References

1. Atlas of the Kyrgyz SSR. T. 1. - M., 1987, – 157 p (in Russian)
2. Dikih AN, The main peculiarities of the conditions of existence of the modern Tien Shan glaciation. Tien Shan glaciation. Frunze: Ilim, S. 3.14 (in Russian)
3. Glaciation of the Tien Shan. - M., VINITI—1995. Responsible editors: Dyurgerov MB, Liu S, Xie Z, - 234 p (in Russian)
4. Glaciation of the Pamir-Alai. - M., Publishing House "Science" - 1993. Executive Editor academician V. M. Kotlyakov (in Russian)
5. Krenke AN (1982) Mass exchange in glacial systems in the USSR. - L.: Gidrometeoizdat, - 114 p (in Russian)
6. Nikonov AA (1982) About the absolute age of the last glaciation in the mountains of Central Asia. In: The development of the nature of the territory of the USSR in the late Pleistocene and Holocene. - M., S. 54–59 (in Russian)
7. Glaciological dictionary. Edited by Corresponding Member of the USSR Academy of Sciences V.M. Kotlyakova. L.: Gidrometeoizdat, 1984, p 528 (in Russian)
8. Imbri D, Imbri K (1988) Mysteries of the ice ages. Traslation from English, Avsyuk GA (ed), M.: Progress, 1988, – 262 p (in Russian)
9. Budyko MI, Butner EK, Vinnikov Kya (1984) Detection of anthropogenic climate change. Meteorol Hydrol 12:5–16 (in Russian)
10. Usubaliev RA, Chen X, Osmonov AT (2013) The geography of glaciation of the mountains of Kyrgyzstan. - c. 135–210. In the book: Physical Geography of Kyrgyzstan. - Bishkek: Turar, – 588 p (in Russian and English)
11. Aizen VB, Kuzmichenok VA, Surazakov AB, Aizen EM (2006) Glacier changes in the central and northern Tien Shan during the last 140 years based on surface and remote-sensing data. Ann Glaciol 43:202–213
12. Dmitry Petrakov, Alyona Shpuntova, Alexandr Aleinikov, Andreas Kääb, Stanislav Kutuzov, Ivan Lavrentiev, Markus Stoffel, Olga Tutubalina, Ryskul Usubaliev. Accelerted glacier shrinkage in the Ak-Shyirak massif, Inner Tien Shan, during 2003–2013. // Science of the Total Envirjnment 562 (2016) 364–378. journal homepage: www.elsevier.com/locate/scitotenv.
13. Kuzmichyenok VA (1991) Fluctuations of glaciers of the Ak-Shyirak ridge from 1943 to 1977. Glaciers-Ocean-Atmosphere interactions. International Association of Hydrological Sciences Publication. - No. 208. - Wallingford, pp 507–513
14. Mandychev AN, Usubaliev RA, Azisov EA (2017) Changes in the Abramov Glacier (Alai Ridge) from 1850 to 2014. Ice Snow 57(3):326–333. https://doi.org/10.15356/2076-6734-2017-3-326-333 (in Russian)
15. Osmonov A, Bolch T, Xi C, Kurban A, Guo W (2013) Glacier characteristics and changes in the Sary-Jaz River Basin (Central Tien Shan, Kyrgyzstan) - 1990–2010. Remote Sens Lett 4(8):725–734. https://doi.org/10.1080/2150704X.2013.789146
16. Podrezov OA, Dikikh AN, Bakirov KB (2001) Instability of climatic conditions and glaciation of the Tien Shan over the past 100 years. Vestn. Kyrgyz-Russian Slavyansk. univ. 2001. - T. 1. - No. 3.- S. 33–40 (in Russian)
17. Solomina O, Barry R, Bodnya M (2004) The retreat of Tien Shan glaciers (Kyrgyzstan) since the Little Ice Age estimated from aerial photographs, lichenometric and historical data. Geogr Ann 86 A (2):205–215
18. Usubaliev RA, Dudashvili AS, Elemanov OI (2012) Glaciation of the northern slopes of the Turkestan and Alai ridges and its modern dynamics. Ice Snow 1(117). - M.: Science. – S. 24–28 (in Russian)
19. The catalog of glaciers in Kyrgyzstan. Bishkek, 2018, 709 S. (https://www.caiag.kg/phocadownload/projects/Catalogue%20%20%20%%ofglaciers%20Kyrgyzstan%202018.pdf) (in Russian and English)
20. Catalog of glaciers of the USSR. Volume 14. Central Asia. Issue 2. Kyrgyzstan. Part 5. Basins of the left tributaries of the Syr Darya River from the mouth of the Aksu River and below. Gidrometeoizdat, Leningrad, 1976, – 91 p (in Russian)

Chapter 8
Assessment of Anthropogenic Load on the Talas River Watershed, Kazakhstan

E. D. Japarkulova and K. E. Kaliyeva

Abstract The analysis of water use features in the Talas River basin was performed. The main indicators of indirect types of anthropogenic impacts on water bodies are considered. The total anthropogenic load within the river basin was estimated based on demographic, agricultural, and industrial components. Agricultural loads are low in the Talas River basin. The average population density in the Talas River basin is 2.76 person/km^2. The level of plowing of the pool on average is 0.79%. The maximum value of the plowing of the territory has low values. The average livestock load in the basin is 0.033 kg goal/km^2, high load is observed throughout the basin. The density of industrial production has an average value of 9.55 thousand dollars/km^2.

Keywords Anthropogenic · Impact · Water bodies · Assessment · Water use · Talas River basin

8.1 Introduction

The objective of this article was to search for reserves to improve the use of water resources based on comparison of total anthropogenic loads with guaranteed water supply of the national economy and natural complexes.

8.1.1 Relevance and Object of Research

The Talas River basin belongs to the Aral Sea basin. The Talas River is mainly formed in Kyrgyzstan in the eponymous intermountain basin. The source of the Talas River is the confluence of the Karakol and Uchkoshoy Rivers at the junction of the Kyrgyz and Talas mountain ranges. Arriving at the plain after traveling through a small gorge, it flows from south to north, splitting into many branches. Further on, in the Sands of Moinkum, the surface flow of water stops and the river is lost in the Sands. The length

E. D. Japarkulova (✉) · K. E. Kaliyeva
Kazakh National Agrarian University, Almaty 050010, Kazakhstan

Z. E. Kulenbekov and B. D. Asanov (eds.), *Water Resource Management in Central Asia and Afghanistan*, Springer Water, https://doi.org/10.1007/978-3-030-68337-5_8

of the river is 661 km, of which 444 km falls within the territory of Kazakhstan. The Talas River basin is divided into two parts—mountain (Taraz) and plain.

Economic activities are a set of factors that cause quantitative and qualitative changes in natural components, and as such, are subject to evaluation and rationing. The urgency of this problem increases due to the growth of negative impacts on the natural environment and the subsequent consequences.

The Talas River basin is located in Zhambyl region, within the sphere of activity of the Shu-Talas basin inspection (BI). Within what is considered the Kazakh part of the Talas River basin, there are six districts of Zhambyl region and the Taraz region, as well as one district of the South Kazakhstan region. The total area of the territory is 1915.1 thousand ha and covers parts of Zhambyl (1772.1 thousand ha), South Kazakhstan (Sozak district 143.0 thousand ha). In general, the Zhambyl region contains approximately 93% of the total area of the basin, and the South Kazakhstan region contains the other approximately 7% [1].

In economic terms, these areas have significant degrees of differentiation from agricultural and industrial to poorly populated and poorly developed (Fig. 8.1).

In this paper, the degree of anthropogenic load on water bodies and on territories of the Talas River basin is estimated on the basis of demographic, agricultural, and industrial components. All areas of the Talas River basin located in the Kazakh part are considered to be the subject of this research.

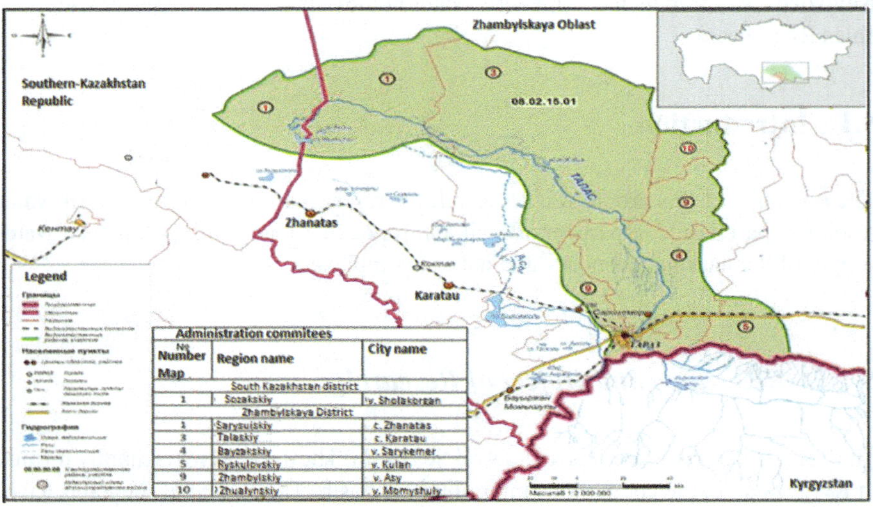

Fig. 8.1 Map-scheme of water management and administrative division of the Talas River basin [2]

8.1.2 Features of Water and Nature Management

The total surface water resources of the territory are under consideration amounts to an average of 808 million m^3 per year. Of these, the water resources form outside of the Republic of Kazakhstan at an annual rate of 716 million m^3 are reduced in low—water years by repeatability 1 time in 4 years ($P = 75\%$) to 637 million m^3, repeatability 1 time in 20 years ($P = 95\%$)—to 529 million m^3. Surface water resources formed in the Republic of Kazakhstan at an annual rate of 92 million m^3 are reduced in low-water years by repeatability 1 time in 4 years ($P = 75\%$) to 72.3 million m^3, repeatability 1 time in 20 years ($P = 95\%$) to 53.5 million m^3. The main source of the rivers of the researched basin is meltwater, specifically the meltwater of seasonal snows. The meltwater of "eternal" snows and glaciers play a significant role in feeding part of the rivers with high-mountain catchments. There are no surface water river tributaries in the Kazakh part of the basin. Therefore, the formation of surface water runoff here is due to the runoff of numerous "Karasu", wedging out channel filtration losses of Talas River water and return irrigation water from irrigation farming basin. It should be noted that the main flow of the Talas River is regulated by the Kirov Reservoir with the long-term goal of regulation being to design volume of filling at the NPU of 520.0 million m^3. The period of intensive filling of the reservoir occurs during the flood season of the Talas River, usually falling between February and April. The territory within Kazakhstan receives runoff according to the Regulations on Water Allocation of Water Resources adopted in 1983. Currently, water allocation limits for vegetation and non-vegetation periods are adopted by the Shu-Talas Interstate Commission [3, 4, 5].

The soil cover of the Talas River basin is characterized by great variety, due to the climatic heterogeneity of the territory, mountain-plain relief, and the presence of groundwater. The diversity of natural and geological conditions causes a complex picture of the distribution of soils and vegetation in the region. It covers a range of landscape zones, from hot and dry deserts to wet Alpine mountain meadows, due to the vertical division of the country, lying in the range from 300 to 4000 m in altitude. Most of the lands having the greatest economic development are located in the areas of low mountains, foothills, and deserts. These areas allow for the division of the pool of soil areas (agricultural areas, as the topography, with geology and moisture conditions determine the possibility of farming, and its systems usage of grassland, etc.).

The soil cover is represented by mountain dark chestnut and light chestnut, gray-earth, light gray-earth, loess and loess loam, meadow-gray-earth, meadow, meadow-marsh, salt marshes, salt flats, takyrovidnymi, and meadow-brown soils. The boundaries of the Talas River basin include the Moynkum sandy area, which occupies up to 30–50% of the basin.

The territory under consideration is considered one of the main industrial and agricultural regions of the Republic, and agriculture significantly affects the entire socioeconomic situation in the region.

The industry of the water management complex of the basin is represented by enterprises of the mining industry and manufacturing industries, producing food products, chemical industry, light, and leather industry.

The total area of agricultural land is 1411.62 thousand ha, including irrigated 55.19 thousand ha. The total fund of developed lands of regular irrigation is 51.06 thousand ha, estuary irrigation 4.13 thousand ha [6].

The population of the territory is 529,248 thousand persons. At the same time, the urban population is about 67.5%, the rural population is about 32.5% [2].

8.2 Research Methods

When characterizing and assessing anthropogenic loads on water bodies, two groups of indicators were taken into account: direct and indirect impact [6]. Direct impacts on water bodies were determined based on the volume of river flow withdrawal.

Indirect impacts on water bodies are manifested in the form of anthropogenic loads on the catchment associated with industrial and agricultural specialization of the economy. Indirect (areal) impacts are important for assessing the intensity of anthropogenic load. As the following parameters were used: population density area (person/km^2); density of industrial production (the volume produced in the region industrial production in thousand USD per 1 km^2) and agricultural development, including plowed (%) and the livestock load (the number of conditional heads of cattle per 1 km^2). The calculations were carried out with reference to the boundaries of the Talas River basin.

Each of these indicators adopted a conventional scale of 8 steps (Table 8.1), which was based on the gradation of the main indicators of anthropogenic load in the author's edition of Isachenko [6]. The applied indicators are grouped by types of anthropogenic impacts: demographic, industrial, and agricultural. The average value of each was estimated as the average level of the corresponding anthropogenic load in the Talas River basin. The agricultural load was obtained as the mean of the score estimates of the intensity of agricultural (plowing) and livestock loads.

8.2.1 Assessment of Anthropogenic Load

The calculations revealed the following features of differentiation of anthropogenic load. The population density within the Talas River basin is an average one (Table 8.2). The average population density in the Talas River basin is 2.76 person/km^2.

Agricultural loads are low in the Talas River basin. The level of plowing of the pool on average is 0.79%. The maximum value of the plowing of the territory has low values. The average livestock load in the basin is 0.033 kg goal/km^2, high load is observed throughout the basin.

Table 8.1 Scale of intensity of anthropogenic load on catchment areas of river basins

Index	The intensity of the load point							
	1	2	3	4	5	6	7	8
	Minor or absent	Very low	Low	Reduced	Average	Elevated	High	Very high
Population density, people/km^2	0.0	<0.10	0.20–1.0	1.10–1.50	5.10–10.00	1.10–25.00	25.10–50.0	>50.0
Density industrial production, thousand Dol/km^2	0.0	<0.35	0.36–3.50	3.60–35.00	36.00–105.0	106.0–140.0	141.0–170.0	>170.0
The plowed, %	0.0	<0.10	0.20–1.0	1.10–1.50	5.10–15.00	15.10–40.0	40.1–60.0	>60.0
Animal load, usl. goal/km^2	0.0	<0.10	0.20–1.0	1.10–2.00	2.10–3.00	3.10–6.0	6.10–10.0	>10.0

Table 8.2 Assessment of anthropogenic load on the catchment area of the Talas river basin

Index	Talas River basin
Total area, thousand km^2	191,51
Population, thousand person	529,248
Population density, person/km^2	2.76
Irrigated land area, thousand hectares	55.19
The plowed, %	0.79
Livestock, thousand heads	632.7
Livestock load, cond. heads/km^2	0.033
Industrial products, thousand dollars	1,828,653,3
Density of industrial production, thousand dollars/km^2	9.55
Available water resources, km^3	0.402
Specific water supply per inhabitant, thousand m^3/person	0.76

The density of industrial production has an average value of 9.55 thousand dollars/km^2. However, the loads associated with industrial production are significantly differentiated by territory and reach their maximum in the middle part of the Talas River, Taraz, where it is estimated as "high." Downstream industrial loads are rated as "reduced."

8.3 Conclusion

The comparative analysis of indirect impact indicators allows for a detailed picture of the total anthropogenic loads, to outline the directions of activities to reduce them. In general, when assessing the intensity of anthropogenic loads, the comparison of relative and specific indicators (such as population density and industrial production, livestock load, plowing of the territory, etc.) increases the objectivity of the results obtained. It allows for the identification of territorial formation patterns and of the functioning of water use systems. Assessment and regulation of anthropogenic loads will allow for the development and proposal of a system of compensatory measures within river basins.

References

1. The basins of the rivers of river Issyk-Kul and rivers Chu, Talas, Tarim. Surface Water Resour USSR 14(2). 2 (1967, 1973, 1977)
2. Scheme of integrated use and protection of water resources of the Talas River basin. Kazgiprovodkhoz, Almaty (2015)

3. Regulations on the division of the flow in the Talas River, Moscow (1983)
4. Annual reports of Shu-Talas Basin Inspection (2006–2014)
5. Materials of the Commission of the Republic of Kazakhstan and the Kyrgyz Republic on the use of water facilities of interstate use on the rivers Chu and Talas, Astana, Bishkek, Taraz (2000–2019)
6. Isachenko AG (2001) Ecological geography of Russia. Publishing house of St. Petersburg state University, Saint Petersburg, p 328

Part III
Risk Analysis for Water Resources Systems, and Related Issues

Chapter 9
Water Resource Risks to Cotton Agriculture in Uzbekistan: Climate, Policy and Irrigation

M. Brody and B. Eshchanov

Abstract The world's most important environmental issue is global climate change. In Central Asia, and specifically Uzbekistan, perhaps the most significant long-term consequence of warming is water resource problems stemming from the decline of the Central Asian mountain glaciers. The disappearance of the Aral Sea is one of the many environmental issues, although primarily man-made. Other risks to water usage for agriculture include the possibility of returning flows to the Aral Sea Basin for the partial restoration of the sea. Although unlikely, any regional water conflicts would be difficult for a country downstream, like Uzbekistan. This paper takes a broad look at the long-term scale of water and climate issues, and the cost issues that are obstacles to the adoption of efficient irrigation technologies.

Keywords Climate change · Glacial decline · Drip irrigation · Cost · Adaptation

9.1 Introduction

Uzbekistan is a major agricultural producer, both regionally and globally. Agriculture remains an important component of national income, generating about 17.5% of Uzbekistan's GDP and affecting the well-being of all the nearly 14.5 million people who live in rural areas. Traditionally, cotton and wheat are the dominant crops in Uzbekistan and they currently take up the dominant share of land resources, although the horticulture subsector has been a source of growth as well [1]. The total water footprint for agriculture in Uzbekistan during the period between 1996 and 2005 was calculated to be 38.3 billion m^3/year; compared to 2.77 and 1.2 billion m^3/year for domestic supply and industrial production, respectively [2].

M. Brody (✉)
American University, 4400 Massachusetts Ave, NW, Washington, DC, USA
e-mail: mbrody@american.edu

B. Eshchanov
Westminster International University-Tashkent, 12 Istikbol Str, Tashkent, Uzbekistan
e-mail: b.eshchanov@gmail.com

Z. E. Kulenbekov and B. D. Asanov (eds.), *Water Resource Management in Central Asia and Afghanistan*, Springer Water, https://doi.org/10.1007/978-3-030-68337-5_9

Future water availability for agriculture in Uzbekistan has a number of serious uncertainties regarding issues with climate, technology, and policy debates. Traditional irrigation systems use enormous quantities of water and have created problems with waterlogging, salinity, and inefficient water usage [3]. The nearly complete diversion of the Amu Darya and Syr Darya Rivers for cotton irrigation has led to the almost entire disappearance of the Aral Sea. Any attempt at a restoration of the Aral Sea will require restoring the former flows there. Although unlikely, regional water conflicts are conceivable, as Uzbekistan is a downstream country. The most significant and difficult aspect to forecast is the long-term effects of the melting mountain glaciers, which, along with annual snow melt are the main water resources in the region.

9.1.1 Climate Change and the Long-Term Issue of Water Availability

This section is a brief review of the climatic issues in Central Asia that will increase the uncertainties in water availability for agriculture in Uzbekistan. First, the average global temperature has increased by about 0.8 °C since 1880. Two-thirds of the warming has occurred since 1975, at a rate of roughly 0.15–0.20 °C per decade [4]. In Central Asia, one of the consequences of this warming can be seen in the losses of glacial area in the Altai-Sayan, Pamir and Tien Shan. Remote sensing data analysis from the 1960s through 2008 informs scientists of this (Fig. 9.1).

More recently, the decline in glacial area in the Pamir Mountains can be seen in Fig. 9.2, and although no quantitative analysis is shown here, the decline in the glaciers over a period of 22 years is striking and should be noted.

In Uzbekistan itself, similar phenomena are occurring, and there is occurrence of both increased temperatures and declining glacial water resources. Average annual temperatures in Uzbekistan are increasing in all regions across the country, as seen here in Fig. 9.3. And in Fig. 9.4, one can see how significantly the number of days with air temperatures over 40 °C has increased [7].

Sources of water resources are rapidly declining, and increasing temperatures are creating higher evaporation/transpiration rates during the agricultural growing seasons.

9.2 Agriculture and Irrigation in Uzbekistan

Cotton is the dominant crop in the agricultural production systems of Uzbekistan. Uzbekistan is the highest agricultural water consumer in Central Asia with an estimated irrigated area of about 4.3 million ha. Farmers still predominantly use outdated

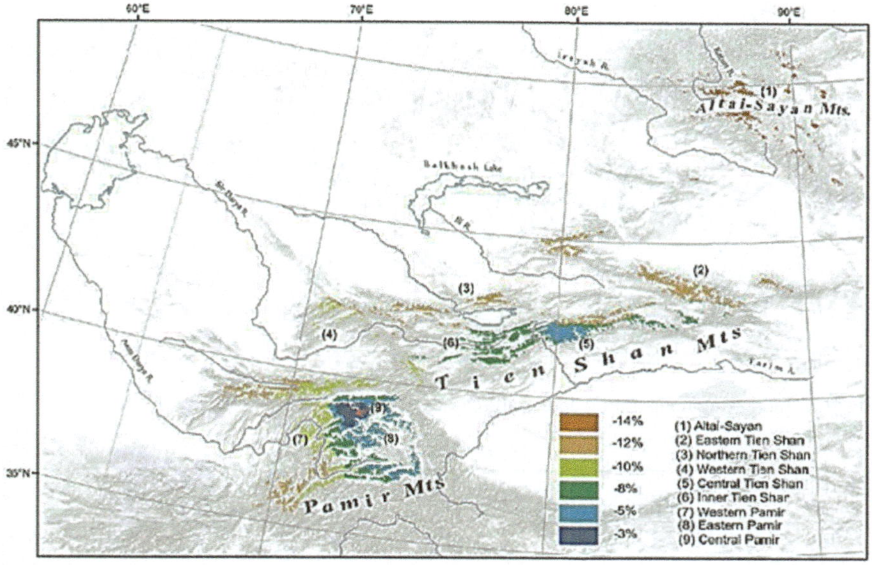

Fig. 9.1 Decline in glacial extent in the Pamir and Tien Shan Mountains [5]

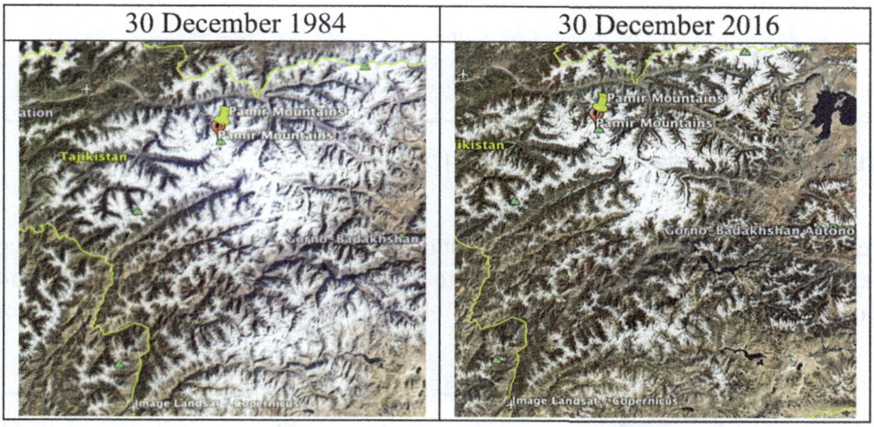

Fig. 9.2 Glacial loss in the Pamir Mountains, 1984 and 2016 [6]

conventional irrigation methods for major crops like cotton and wheat. These inefficiently irrigated agricultural practices remain an important economic activity in the region's arid and semi-arid climates. Full-control irrigation, sprinkler irrigation, and drip irrigation are applied in over 90% of the region's irrigated areas [8], where water usage is inefficient due to outdated irrigation infrastructures [3].

For the 2016/2017 production year, 1.18 million ha of cotton were planted in Uzbekistan [9]. This number has declined very slightly as the country is trying to

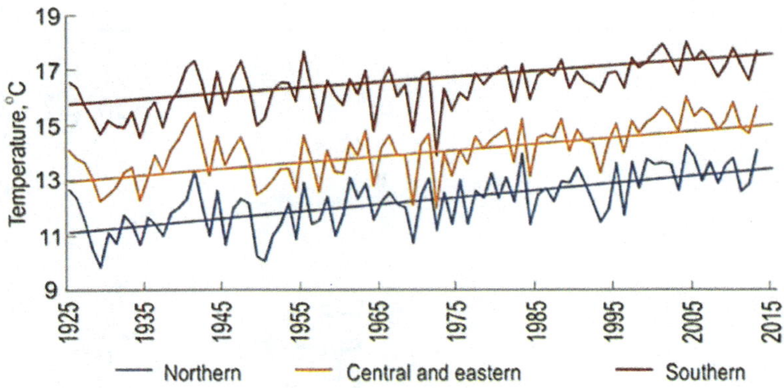

Fig. 9.3 Increasing temperature across all regions of Uzbekistan [7]

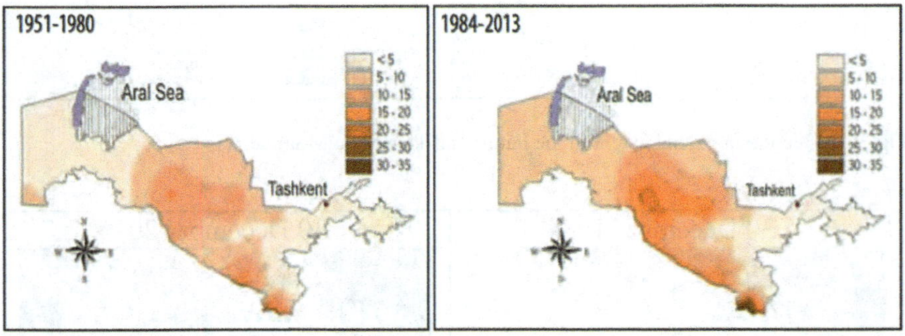

Fig. 9.4 Average number of days above 40 °C [7]

reduce its economic dependence upon cotton and increase wheat production for food security. Uzbekistan is also attempting to develop markets for more valuable fruit and vegetable orchard crops. Figure 9.5 [10] shows historical production levels of

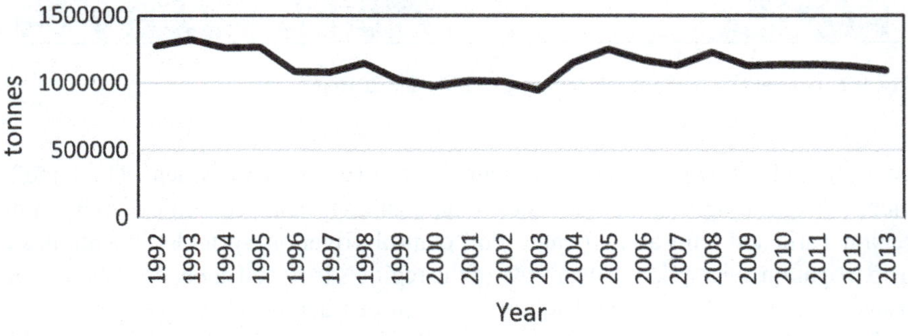

Fig. 9.5 Cotton lint production [10]

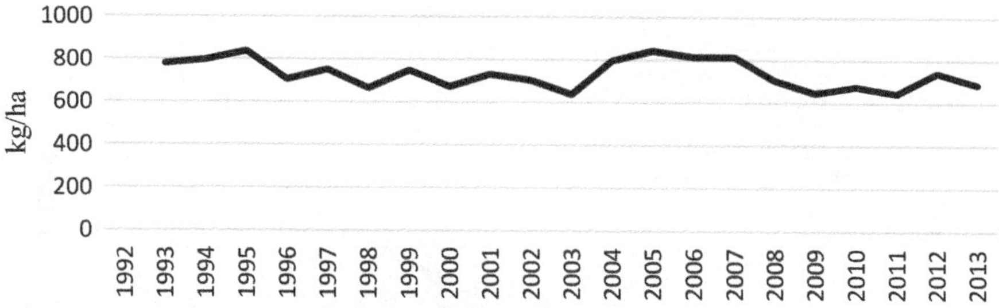

Fig. 9.6 Cotton yield [11]

cotton lint. The difference between the maximum and minimum production years was 28%. Figure 9.6 [11] shows cotton yield per hectare. The difference between maximum and minimum yield was 25%. Whether any of this variation can currently be attributed to annual water availability is unclear.

9.2.1 Future Warming and Its Potential Consequences

It is not possible to predict the exact magnitude of the effect that the warming of Earth will have due to increasing greenhouse gas concentrations in the atmosphere. This is mostly due to the fact that we do not know if countries of the world will successfully control or seriously reduce total global greenhouse gas emissions. This is a major policy uncertainty. The IPCC developed four scenarios of the annual emissions concentration pathways to model the future of climate change. These scenarios which are referred to as "representative concentration pathways" or RCPs [12] range from significant progress in decarbonizing the world's economy to scenarios with little or no control over future emissions, and therefore very high increase in greenhouse gas concentrations in the atmosphere.

According to the IPCC, "warming trends and increasing temperature extremes have been observed across most of the Asian region over the past century" [5]. Water scarcity issues are likely to become a very serious issue. This is in part due to human factors, such as an increasing demand and a lack of good water resource management. The potential impacts on food production will vary by region. In Uzbekistan, the possibility of more frequent droughts could negatively affect cotton production, increase water demand for irrigation, and worsen the existing land quality issues (IPCC 2014). Potential summer warming in Central Asia is shown in Fig. 9.7 [12].

The consequences relating to water resources in Uzbekistan can be seen in Fig. 9.8. There are pre-recorded declines and forecasted declines in glacial volume in four Uzbek river systems as shown below [7].

This is similar to what was seen on a regional scale with Central Asian mountain glaciers (Figs. 9.1 and 9.2), but on a smaller geographic scale. The possible consequences of this warming and water resources pose risks to three major crops

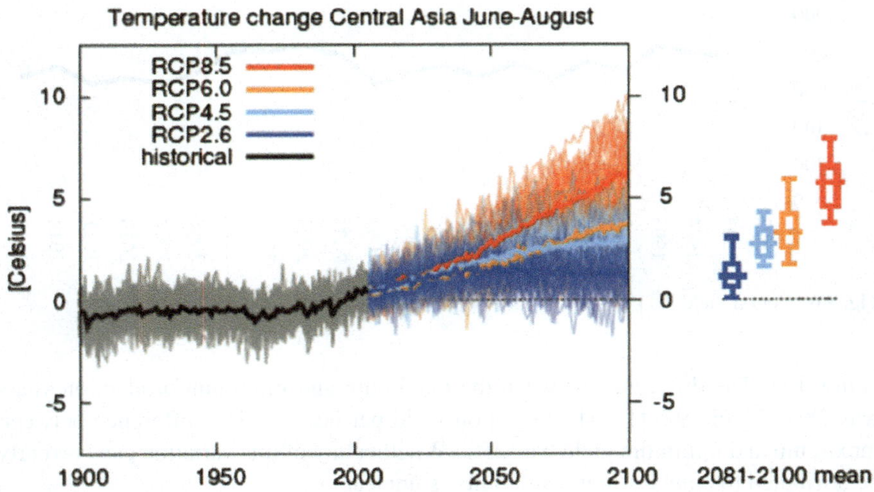

Fig. 9.7 Scenarios of summer warming in Central Asia [12]

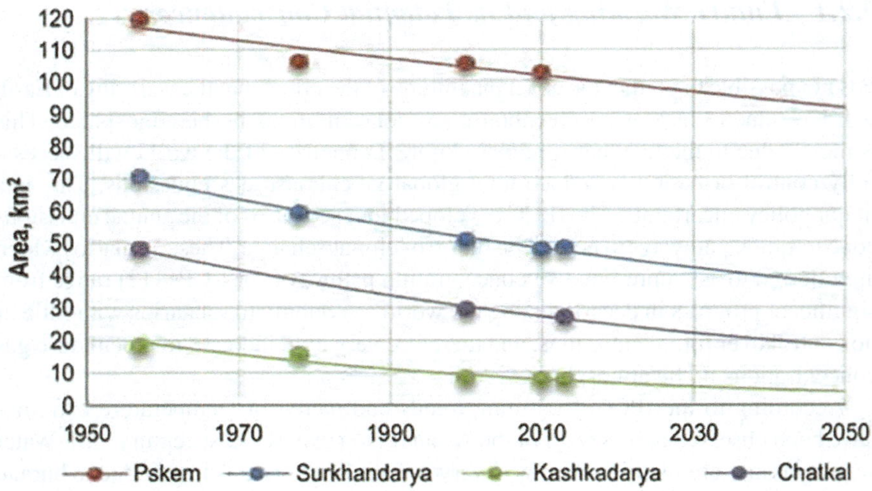

Fig. 9.8 Observed and forecasted decline in glacial area in 4 Uzbek River Basins [7]

in Uzbekistan (Table 9.1). Under these three warming scenarios, the potential crop losses range from more modest losses to potentially 50% in some crops under the high warming scenario.

Table 9.1 Forecasted crop loss under three scenarios of climate change (in %) [7]

	Desert and Eastern Steppe			Desert and Western Steppe			Piedmonts Southwestern part			Piedmonts Eastern part		
	L	M	H	L	M	H	L	M	H	L	M	H
Cotton	−11	−10	−36	−19	−20	−31	−16	−21	−32	−3	−17	−49
Fruits	−13	−12	−49	−23	−22	−39	−20	−25	−42	0	−18	−63
Winter wheat	−1	−2	−40	−13	−20	−32	−17	−21	−43	−19	−7	−42

9.3 Discussion

Adapting to climate change in Uzbekistan requires a significant reduction of wasted irrigation water, and that can only be done in a limited number of ways. The major irrigation water transportation canals could be lined with plastic sheeting to significantly reduce ground infiltration. This could also increase the efficiency of these canals by almost 40% [13, 14]. Although extremely labor intensive, approximately 2500 km of these transportation canals throughout Uzbekistan could be lined for a cost of approximately $40 million, which, considering the benefits, is not an obscenely large cost [14].

Another method to greatly improve irrigation efficiency in Uzbekistan would be to only use drip irrigation technologies for cotton production. A range of estimates can be found in published literature. In 1997, the reported average cost of irrigation development was about $11,200 per hectare for surface irrigation schemes using standard modern technologies, including agricultural infrastructure. Rehabilitation and modernization costs of the old irrigated areas would potentially cost an estimated $4500 per hectare. The two main elements of this work would be land leveling and the introduction of modern irrigation techniques. The cost of drip irrigation development on existing irrigated areas varies between $2300 and $3500 per hectare [8]. Other cost estimates range from $3000 per hectare to $5000 per hectare [1]. Drip irrigation not only saves water, but it also typically improves crop yields at the same time (Table 9.2). Various studies have reported water savings of 20–40% when using drip irrigation [3]. The fertilizer and energy savings can also be significant [15].

Moving to drip irrigation would bring significant benefits in yield along with a reduction in water use. With the current 1.18 million ha devoted to cotton production, conversion of all of this area to drip irrigation could potentially reach installation costs of $6 billion, perhaps even higher. It is important to understand that there is significant uncertainty with this number, and that some of this land is already salinized for drip irrigation, but these high costs are the primary reason that despite the many research projects in this area on irrigation improvements, there has not been a wholesale adoption of these technologies.

Table 9.2 Comparison of cotton yield between traditional and drip irrigation systems in the Karshi Steppe, in Uzbekistan [3]

Technology	Yield (kg/ha)
Traditional irrigation system	3508
Drip irrigation	4598

9.4 Conclusion

Uzbekistan, currently, has an abundance of available water. But it is a downstream country in the overarching Central Asian hydrologic system. The effects of climate change are already being observed with the decline of water sources and a general rise in temperatures. These two factors breed the resulting stresses on agriculture. Cotton, while still very important to the economy, is an extremely water-intensive crop.

There are many uncertainties about future warming; however, there is little doubt that the agricultural stresses will increase with the rising temperatures. The only question is by how much, and that question cannot be answered, at least not yet. The data on the costs of modern irrigation systems should be considered somewhat preliminary, until serious attempts at scale-ups have been made. Improving irrigation will be a fundamental aspect of Uzbekistan's adaptation to future water resource uncertainties. Furthermore, should the country ever consider returning any quantity of flows to the Aral Sea, then these irrigation technologies will become even more important.

The scale of investment required for upgrading to modern irrigation technologies is undeniably very large. So, fundamental economic questions about any investment strategies will be how much, how quickly, where, and who pays. These questions are a combination of economic and policy questions and will not be answered by simply continuing with more small-scale research projects. One approach to this economic analysis could be real options analysis, but would be the subject of yet another analysis and is beyond the scope of this paper [16].

References

1. Larson DF, Khidirov D, Schuman I (2015) Uzbekistan—strengthening the horticulture value chain (English). Uzbekistan vision 2030 background paper series. World Bank Group, Washington, DC. https://documents.worldbank.org/curated/en/396111468301526337/Uzbekistan-Strengthening-the-horticulture-value-chain
2. Hoekstra A (2013) The water footprint of modern consumer society. Routledge, London. 224 pages. https://doi.org/10.4324/9780203126585. ISBN: 9781136457050
3. Djumaboev K, Manthrithilake H, Ayars J, Yuldashev T, Akramov B, Karshiev R, Eshmuratov D (2019) Growing cotton in Karshi Steppe, Uzbekistan: water productivity differences with three different methods of irrigation. In: Conference: 9th international micro irrigation conference (9IMIC). Indian National Committee on Surface Water (INCSW)—CWC Ambassador Ajanta, Aurangabad, India 16–18 Jan 2019
4. NASA Earth Observatory, https://earthobservatory.nasa.gov/world-of-change/DecadalTemp. Accessed on 4 Apr 2019
5. IPCC (Intergovernmental Panel on Climate Change) (2014) Assessment Report 5, working group 2. Impacts, adaptation & vulnerability. https://www.ipcc.ch/pdf/assessment-report/ar5/wg2/WGIIAR5-Chap24_FINAL.pdf
6. Landsat photography, Google earth. Accessed on 4 Apr 2019

7. UzHydromet (Uzbek Hydrometeorological Service) (2016) Third national communication of Uzbekistan under the UN framework convention on climate change. https://unfccc.int/national_reports/non-annex_i_natcom/items/10124.php
8. FAO (Food and Agriculture Organization of the United Nations) (2013) Irrigation in Central Asia in figures: AQUASTAT Survey-2012. FAO Water Reports 39. Available from: https://www.fao.org/3/a-i3289e.pdf
9. USDA (United States Department of Agriculture) (2019) World agricultural production. Foreign Agricultural Service. Circular Series WAP 3–19 Mar 2019
10. FAOSTAT (2019) https://www.fao.org/faostat/en/#data. Accessed on 8 Mar 2019
11. Index Mundi (2019) (Data from the USDA). Accessed on 4 Apr 2019. https://www.indexmundi.com/agriculture/?country=uz&commodity=cotton&graph=yield
12. IPCC (Intergovernmental Panel on Climate Change) (2013). Assessment report 5, working group 1. Climate change: The physical science basis; Summary for policy makers. https://www.ipcc.ch/site/assets/uploads/2018/02/WG1AR5_SPM_FINAL.pdf
13. Hojyaz J, Eshchanov R, Nurmetov J, Lutz A, Shanafield M, Lamers J (2013) Canal lining to increase water use efficiency and remediate groundwater levels in Khorezm Uzbekistan, Central Asia. Int J Agric Res Rev 3(4):742–750, Available online at https://www.ecisi.com. ISSN 2228–7973 ©2013 ECISI Journals
14. Rudenko I, Salaev S, Davletov S, Eshchanov R (2015) Economic feasibility study of canal plastic lining in the Aral Sea basin. In: Lee TS (ed) Chapter 10 in: Research and practices in water quality, InTech. http://dx.doi.org/https://doi.org/10.5772/59101
15. UNDP/GEF (2013) Newsletter #4. Land energy biodiversity. Global environment facility's small grants programme 100015, 41/3Mirabad str., Tashkent (www.sgp.uz)
16. Golub A, Brody M (2017) Uncertainty, climate change, and irreversible environmental effects: application of real options to environmental benefit-cost analysis. J Environ Stud Sci 2017(7):519–526. https://doi.org/10.1007/s13412-017-0436-7

Chapter 10
Groundwater Flooding Risk Assessment Using Microseismic Arrays and VES Techniques in the Northern Part of Bishkek

Sagynbek Zh. Orunbaev, Baktyiar D. Asanov, and Zheenbek E. Kulenbekov

Abstract Increased groundwater elevation and subsequent flooding of infrastructures are a threat to Bishkek city. Novel geophysical approaches to determine and map groundwater depths, such as microseismic arrays and VES technologies, were used in the northern part of Bishkek. These new geophysical measurements showed groundwater at a depth of 5–10 m at all points. This is considerably deeper than the 1–3 m currently shown on flood risk maps. Such discrepancy (2–7 m) is due to seasonal water level changes observed at approximately 5 meters which were not captured in past monitoring. The lower water levels measured in this study also indicate the impressive performance of the collector-drainage systems, filter channels, and tunnels, since the geophysical measurements were carried out in the wet spring and summer seasons. The study showed how geophysical measurements can be used in groundwater mapping and flood risk assessment.

Keywords Geophysical measurements · Groundwater · Floods · Central asia

10.1 Introduction

Groundwater flooding is a phenomenon that occurs when the groundwater level rises to fill infrastructure and housing foundations. Mapping of groundwater floods is not done much yet on a global scale. In Kyrgyzstan, groundwater flooding can also lead to slope geotechnical failure which is hazardous. The Chui Oblast territory is the most affected by flooding. It is there where these processes are developed with an area of about 1,700 km^2 along the natural nip line.

As a consequence, weakened soils form under the influence of groundwater. The main displacements and deformations of earth which occur on these zones are accompanied by the destruction of housing stock and urban infrastructure. A high level of groundwater has a significant impact on buildings and structures due

S. Zh. Orunbaev (✉) · B. D. Asanov · Z. E. Kulenbekov
American University of Central Asia, 7/6 Aaly Tokombaev Str, Bishkek, Kyrgyzstan
e-mail: orunbaev_s@auca.kg

Z. E. Kulenbekov and B. D. Asanov (eds.), *Water Resource Management in Central Asia and Afghanistan*, Springer Water, https://doi.org/10.1007/978-3-030-68337-5_10

to the deep penetration of foundations and underground facilities. Thus, appropriate mitigation measures should be taken during construction in accordance with the hydro-geographical area of development. Unfortunately, the construction of housing without appropriate measures being taken to prevent these gross violations of construction norms and rules remains the de facto practice in the area [13]. Most developers have been shown to ignore the requirements of the resolutions enacted by the district administrations. These resolutions prohibit any construction on the allocated land plots without obtaining the relevant project documentation and approval. For example, in the United States, the Recommended Seismic Provisions for New Buildings and Other Structures by the National Earthquake Hazards Reduction Program (NEHRP) serves to ensure safe construction of infrastructure and household structures in zones at risk of seismic activity [5].

This study used two relatively new technologies, vertical electric sounding (VES) and microseismic array sounding, which were previously not available in Central Asia to update current knowledge on flood risk in an expanding and flood-prone region of the city of Bishkek. The use of VES and seismic profiling as a geophysical prospecting technique is relatively inexpensive and it generates accurate hydrogeological measurements. The purpose of VES is to obtain a general idea of the structure of the geoelectric sections at sites.

10.2 Fieldwork Data

10.2.1 Study Area

The study was carried out in a residential area of Bishkek, the capital of the Kyrgyz Republic. In the study region, previous surveys indicate that there is a high risk of flooding by groundwater posing risks to urban construction and foundations (Fig. 10.1). There are 92 settlements, including Bishkek, Kant, Tokmok, and Kara-Balta in the flood zone [3].

Flooding occurs in the northern part of the city of Bishkek, north of the Big Chui Canal (BCC). This area is characterized by the sinking of the earth with a high water table. The depth of the groundwater does not exceed 3 m in 80% of the area of this zone. Groundwater levels on the surface largely depend on the conditions of the collector-drainage systems (CDS). The condition of the CDS is important, as its purpose is to divert water into run-off channels [3].

10.2.2 Fieldwork Data

Data collection was carried out at all four sites over two periods of fieldwork, one in spring (May 2018) and one in summer (June 2018). Groundwater level (GWL)

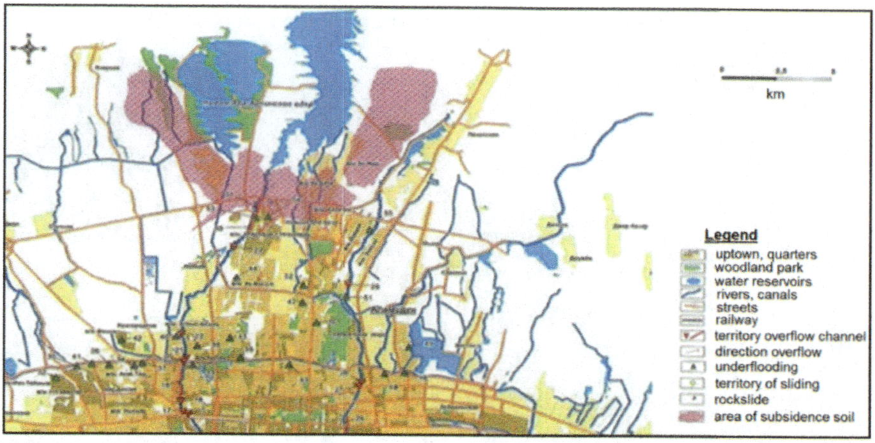

Fig. 10.1 Map displaying areas of urban development and hazardous natural processes in the north part of Bishkek (MES report, 2014)

observations were performed from May 01 to June 15 of 2018. Four sites were identified for exploration (Fig. 10.2). These points are characterized by a complex geological structure and were determined to have hydrogeological zonality inherent in all alluvial cones. The head and middle parts of the alluvial cone are the loci for the main area of groundwater accumulation. Here, the source of power are filtration losses from both the BCC and the nearby Komsomolskoye Lake. As a result, an aquifer complex with a free surface of 100–190 m has been formed (the highest values are indicative for the middle part of the cone).

The depth of occurrence of the aquifer system decreases in the northern direction from 55 m to the first meters on the southern boundary of the wedge zone where the water return is 15–23%. The amplitude of level fluctuation reaches 0.5–2 m south of the BCC according to data from the periodic observations conducted by the MES KR [3].

GW1, GW2, GW3 and GW4 investigated sites shown in Fig. 10.2. The first well was established at the Komsomolskoe Lake (GW1), where the GWL is close to the surface. The other three soundings were arranged in a northwestern direction near the Big Chui Canal (BCC). GW2 was positioned in the 110-ward, then moving northwards GW3 was positioned, near Nijnaia Maevka. Lastly, GW4 was established near the central part of the north side of the city, Karagachevaya Rosha. A total of four VES profiles were obtained for two seasons: spring and summer.

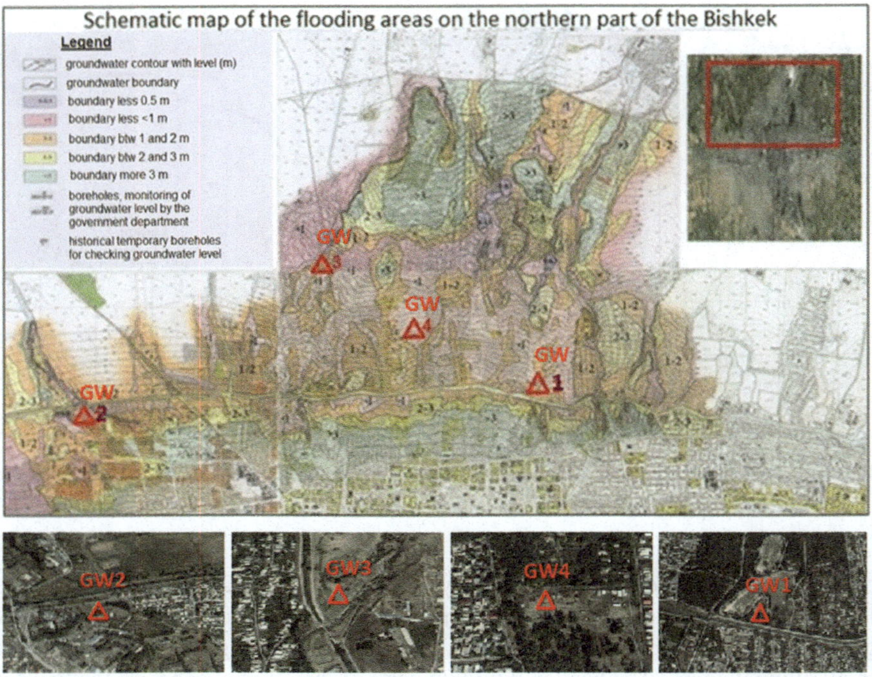

Fig. 10.2 Map of groundwater flooding in Bishkek and location of experimental fieldwork (red triangles indicate areas for conduction of seismological and vertical electrical soundings)

10.3 Methodology

10.3.1 Methods

The effectiveness of the application of geophysical methods is important. We used two different physical methods: seismic and electrical. The VES and seismic array (SA) methods were used in the study. The reliability of the obtained results depends, to a large extent, on the correct choice of methodology for the geophysical field conditions, especially in the search for and the exploration of groundwater.

10.3.2 Vertical Electrical Sounding

We used the LandMapper ERM-02, the newest device among the licensed Landviser, LLC (Penza, Russia) devices [10, 11]. This portable device measures three important electrical properties in solid, semi-solid, and liquid environments: electrical resistance (ER), electrical conductivity (EC), and electrical potential (EP).

Electrical exploration using the VES method was used to search for pore-porous waters in the loose sediments of the lower part of the BCC and in the river valleys. During both the first and second stages of data collection, the sites were selected according to the results of reconnaissance geophysical studies carried out by the authors via their literature review of the extant hydrogeological survey data.

The Land Mapper ERM-02 was utilized to acquire the VES data from the profiles. The spacing of the electrode probes was set at 2.00, 4.00, 6.00, …, 20.00, 22.00, 24.00, …, 40.00, 44.00, 46.00 m, where the depth of each electrode was set at approximately 0.15 m with the electrode spacing measured at each point. The VES data for each month was measured within two days. This VES data was processed and interpreted by a method of curve matching using the RES1D software [12]. All data was inverted with an average root mean square (RMS) error of less than 5% in each VES profile.

The VES fieldwork began with the implementation of a small amount of reconnaissance VES. The results allowed for the study of the main elements of the relief (river beds, floodplains, terraces, slopes, and watersheds) and areas within the reconnaissance limits. The areas of study differ sharply in terms of geological and tectonic conditions, the nature of water manifestations, and so on. The number of reconnaissance soundings at the research site was 23 points in each of the four sites. Past literature and stock material available for the area of research such as survey and drilling work data were used to select the specific points for VES affixation.

In field studies, significant distortions of VES curves can indicate inhomogeneities associated with technogenic processes that develop due to the disturbance of vegetation and soil cover and/or changes in temperature and humidity conditions near mining excavations. It is effective to perform parametric VES at small distances from mining projects, outside of sites with disturbed surface conditions for this reason (usually 5–10 m from the mouths of wells and pits). It is necessary to set up crosses and either circular, symmetrical, or bilateral three-pole VES to crosscheck results with two components in order to detect possible lateral influences.

The largest volume of the VES work was conducted after preliminary processing of reconnaissance and parametric sounding data, and following the on-site survey of the microseismic array data. Detailed work on the VES method was confined to sites that were promising concerning groundwater levels according to the seismic survey methods. The remaining probing profiles were performed on a smaller network. The minimum distance between VES in the most difficult and variegated conditions was 10 m. The maximum distances under relatively homogeneous geological conditions were approximately 200 + m. In our case, the distance was 92 m.

10.3.3 Microseismic Surveys

Measurements were made in four different areas of the city using 16 stations over the course of two hours in order to obtain information on local velocity structures. Records were digitized at a rate of 100 samples per second and the time series was divided into 60 rows. A configuration of the seismic network is shown in Table 10.1.

Table 10.1 The configurations of the registration network used for conducting field measurements using the MS array [6, 7, 14]

N	Description	Type of array
1	Shooting with 4 seismic sensors and 2 different interoffice distances	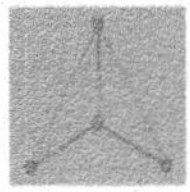
2	Shooting with 7 seismic sensors and 5 different interoffice distances	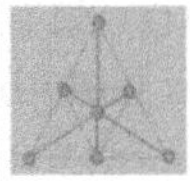
3	Shooting with 10 seismic sensors and 8 different interoffice distances	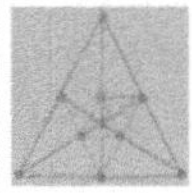

10.4 Analysis and Interpretation

10.4.1 Methodology of Interpretation of VES Curves

The VES method is a geophysical method used to search for groundwater. Despite the difficulties of interpretation, as discussed below, it is so far the only method that allows a quantitative interpretation of the results of observations under favorable geoelectric conditions. It is possible to determine the depth of occurrence of the roof and the base of frozen rocks as well as the depth of occurrence of groundwater, and to approximately estimate their mineralization with the help of this method. Correctly selecting fieldwork methodology suited to the location and the inclusion of distorting factors in the processing and interpretation of field materials improves the effectiveness of the VES method.

The main factors that complicate the interpretation of the VES curves in the upper part of the section are: the variability of the specific electrical resistance (SER), in terms of area and depth due to changes in composition and temperature of rocks and seasonal change in the parameters of the geoelectric section, and types of VES curves. Thus, the geoelectrical complexity of the zones of tectonic faults and contacts of rocks of different petrographic composition needs to be considered.

The processing of field materials began with the grouping of the VES curves on large geomorphological elements (river valleys, root slopes, watersheds, etc.). The types of curves were divided into small elements due to their variety (beds, low or high floodplains, terraces, etc.). Further grouping of the curves was carried out if the characteristics of the curves changed significantly with each element. A qualitative analysis was performed and the types of curves and geoelectric sections were established. The observed forms and types of VES curves often conform to the lateral influence of vertical and inclined boundaries of frozen and thawed rocks, irregularities in the roof of frozen rocks, gradient changes in resistance over depth, and do not always correspond to the real structure of the geoelectric section [4]. All this data was taken into account when analyzing the curves.

10.4.2 Methodology of Interpretation of Seismic Array Measurements

The measurement of seismic noise in local areas through changes in shear wave velocity with depth is a key factor in assessing seismic hazards and is widely used in seismology engineering. It provides information on local effects on the soil and is important for characterizing soil properties.

Seismic noises (microseisms) obtained on area measurements are composed by and generalized as Rayleigh and Love waves, these are analyzed in order to obtain information on the structural velocity of the S-wave in the investigated area. Several methods for obtaining information on the basis of seismic noise records and on the speed of propagation of seismic waves in the sedimentary layer are described in the literature [1, 2, 6, 7, 14]. There were four seismic networks installed in the four research locations in the northern part of Bishkek.

Some of the figures presented in this manuscript have been generated using this GMT software package.

10.5 Results and Discussions

10.5.1 Methodology of Interpretation of VES Curves

Four wells were analyzed to assess the reliability of the VES method in determining the groundwater level (GWL) for two seasons, spring and summer (Table 10.2). Based on the results and interpretation of the VES data collected and the soil layer points at GW1, the depths of the GWLs ranged from 4.0 to 5.6 m during the studied period. The corresponding GWLs when measured seasonally ranged between 3.7 and 6.3 m (Fig. 10.3).

Figure 10.3a depicts a classical 3-layer curve, which shows a decrease in resistance with increasing depth. The resistance decreases by 4–6 m and then slightly increases at the depth of 9–12 m, then again decreases with increasing depth. These can be natural layers or artificial disturbances of the cover of the upper layer. This layer can be very dry and sandy-stony.

At the GW2 site, the depths of the GWLs ranged from 7.3 to 8.5 m during the studied period. The corresponding GWLs when measured seasonally ranged between 6.7 and 8.3 m (Fig. 10.3b).

The measured GWLs were deeper than the estimated GWLs from the map (Fig. 10.2) at the GW2 site. The differences between the measured and estimated

Table 10.2 The characteristics of a geoelectrical section on the investigated area GW1

Results of inversion by RES1DINV software				
Iteration 8				
Layer	Resistivity, (Ohm)	Thickness, (m)	Depth, (m)	Resistivity, (Ohm)
1	714.301	1.02	1.02	714.301
2	916.935	1.275	2.295	916.935
3	15.252	1.594	3.889	15.252
4	34.716	1.992	5.881	34.716
5	142.408	2.49	8.371	142.408
6	67.361	3.113	11.484	67.361
7	15.694	3.891	15.375	15.694
8	7.171	4.864	20.239	7.171
9	11.525			11.525

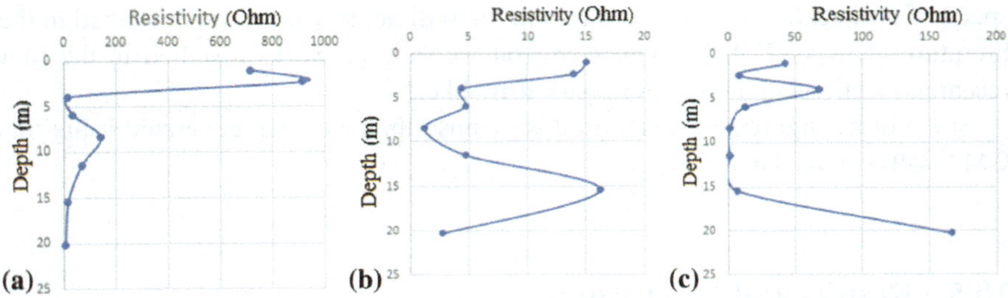

Fig. 10.3 The results of measurements on the distribution of resistivity with depth: **a** at the site GW1, May 2018, **b** in the section GW2, May 2018, **c** at the site GW3, May 2018

measurements ranged from 6 to 7 m, which were within the range of capillary rise in the studied area. The low resistance values (<1 O), at about 8 m, indicate the salinization of the surface layer, where the resistance (10–15 O) is characterized as clayey well-moistened soil according to the results shown in Fig. 10.4b.

For area GW3, the results and interpretation of the VES data point to the depth of GWLs which ranged from 2.3 to 3.5 m during the studied period (Fig. 10.3c). The corresponding GWLs measured seasonally ranged between 1.7 and 3.3 m.

The VES method was used to detect salinity. At the level between 2 and 3 m of the vertical profile in GW3 appears GW and surface part to be saline (Fig. 10.3c) for spring measurements. This site was taken for experimental measurements since Nijnaya Mayevka is sufficiently isolated enough from the problematic areas below the BCC. This lateral influence of thawed rocks is observed under probing in the vicinity of continuous permafrost.

The following seismic velocity profiles are determined by seismic survey data on the Komsomolskoye Lake territory (GW1, Fig. 10.4), an area of 110 m^2, (GW2, Fig. 10.4) and Karagachevaya Roschya (GW4, Fig. 10.4).

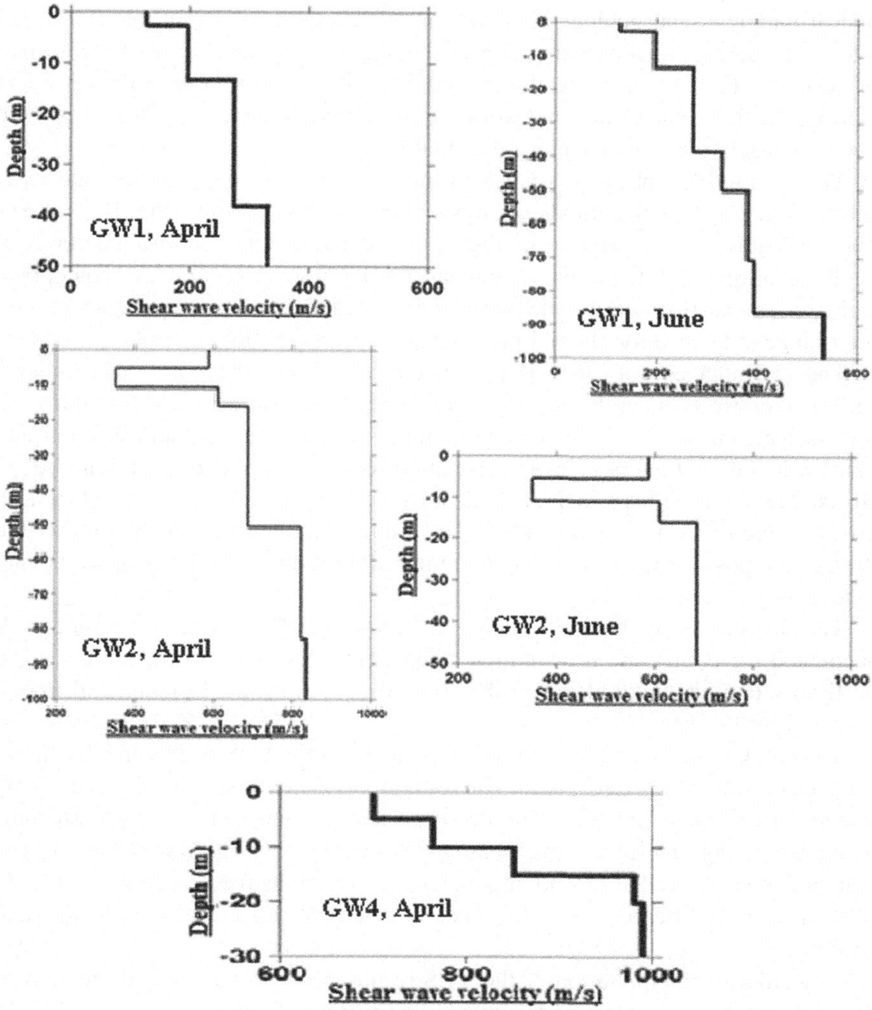

Fig. 10.4 The results of seismic survey measurements on the GW1, GW2, and GW4

10.6 Results and Discussions

In order to assess the risks and dangers associated with further development northward in Bishkek, we identified areas with the highest level of probability of flooding for subsequent geophysical assessment. We then carried out measurements, compared the results of two different methods (VES and MS) for determining groundwater levels, the interpolation of shear wave velocity for the upper 30 m for other parts of the region, and assessed the functionality of the collector-drainage systems (CDS) in place. It is desirous that the full consideration of seismic, geodynamic, hydrogeological, and geological conditions within the territory in this study will allow for more rational planning concerning the development of the northern part of Bishkek. This research has the potential to influence the new buildings being constructed there, since many unfavorable plots are being allocated for development due to the lack of alternative plots of available land.

The complex geophysical studies in the studied areas were carried out using seismic array and VES methods. Analysis of the velocity profiles (Fig. 10.4) shows that the lithological boundaries are distinguished and that the sections are homogeneous according to different measurements. Furthermore, according to the geophysical data, the sections on the measurement points have resistance parameters with their range indicative for alluvial-pro-alluvial deposits and loessal soils.

The advantages of the array method include the fact that microseisms/microtremors can be recorded anywhere. When surface waves dominate in microseisms/microtremors, there is a dominance of the fundamental fashion and plane waves do not intersect (with correlation zero). When the method is applied to Bishkek [8, 9], it builds a map of shear wave velocity for the city, using slope as a proxy. Therefore, it is recommended to record results for more than 30 min so that there is the possibility of excluding temporary interference from nearby traffic and other factors.

The distortions of the VES curves are most significant and often cannot be accounted for if they are near the receiving electrodes. It is not recommended to perform sounding on old glades, paths, roadsides, or along tracks laid by all-terrain vehicles, for this same reason.

The initial material results of the VES method was the VES graphs and the digital material on which they were based. The resulting VES field charts were processed by a standard procedure, including 2 processing steps: qualitative processing and quantitative processing (interpretation). Qualitative processing consisted of comparing the obtained types, as well as of studying spatial patterns of distribution along the profile of regions with different values of apparent electrical resistances on geoelectrical sections.

A qualitative interpretation of the VES results was carried out with the aim of determining the parameters of the geoelectric section, the thicknesses and specific electrical resistances of the horizons composing the section, as well as the depths of their occurrence. The interpretation of each VES observation point was carried out by RES1DINV software. The computer automatically selects a theoretically calculated VES schedule for the greatest coincidence with the VES field chart, according to the program. In this case, the values of the electrical resistance of the selected horizons and their power, according to the chosen schedule (theoretically), are taken as true.

It was very difficult to interpret the VES curves obtained from the GW4 section (Fig. 10.5). The soil is strongly saline throughout the profile rendering interpretation over the layers is meaningless (see Fig. 10.3c).

The geoelectric sections were constructed for each point of the VES (resistance and power) with the correspond horizons and values of their electrical resistances

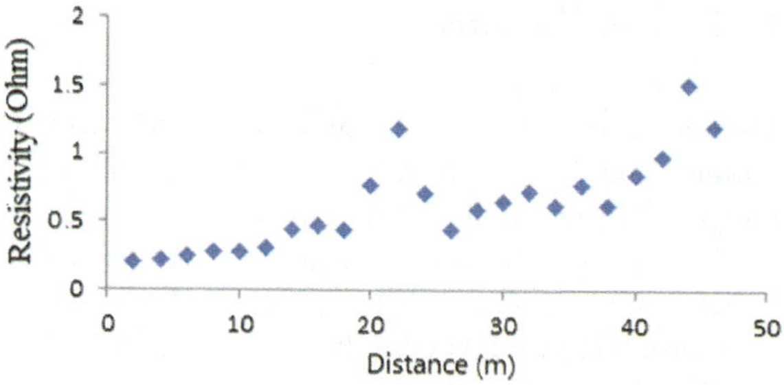

Fig. 10.5 The results of the measurement of the resistivity distribution with depth at section GW4, May 2018

Table 10.3 Overall characteristics of the results of executed work

№ Section	Length of the profile/area, m	Resistance of the soil (КС), ohm*m	Level of ground water (VES), m	Velocity profile, Vs30, m/s	Level of ground water (array), m
GW1	96	5–150	4–6	175	1–4
GW2	96	1–20	7–9	420	5–10
GW3	96	0–70	5–15	–	–
GW4	96	–	–	180	4–8

based on the results of the interpretation. Interpretation of the results of electro-sounding is presented in the form of graphs of VES and geoelectric profiles on the profiles constructed based on these graphs.

The sites selected for the conduction of geophysical measurements were those with the highest probability of flooding based on a groundwater level between 1 and 3 m. The results, however, indicate that the groundwater table is observed on average at a level between 5 and 10 m based on the VES and array data (Tab 10.3).

The differences between the observed and expected values of between 2 and 7 m at the sites can be explained by

(a) fluctuations in groundwater, which can be up to 5 m, and/or;
(b) an excellent level of work by collector-drainage systems (CDS).

As the geophysical measurements were carried out in the spring and summer when the water table is typically higher, the finding that the groundwater was at a depth of 5–10 m likely indicates positive performance by the CDS and filter channels/tunnels. The only exception was at the GW1 site, where the groundwater level was at 1–2 m, possibly indicating the need for a better filtration tunnel.

High groundwater levels not only increase flood risk, but the influence seismic resistance within the soil and buildings increases the likelihood for liquefaction during a large earthquake.

10.7 Conclusions

The results of complex geophysical measurements (seismic array and VES) show groundwater at a depth of 5–10 m at all points where previous flood maps record levels of 1–3 m. High variation in groundwater levels (2–7 m) was observed.

The results of this study indicate the excellent conditions of the collector-drainage systems in place. Groundwater levels at the locations estimated to pose the highest flood risk were calculated to be 5–10 m. This is in contrast to the current literature which indicates a level of 1–3 m. With a gap of 2–7 m it is possible the amplitude variation of groundwater is about 5 m over the long term. However, measurements were carried out in the spring and summer seasons, when groundwater level is the highest following the rainy period, suggesting that the effectiveness of the CDS played a role in regulating the level.

This study carried out this work with the support of the United States Agency for International Development (USAID).

References

1. Aki K (1965) A Note on the Use of Microseisms in Determining the Shallow Structures of the earth's Crust. Geophysics 30:665–666
2. Aki K (1957) Space and time spectra of stationary stochastic waves, with special reference to microtremor. Bull Earthq Res Inst Univ Tokyo 35:415–457
3. Ministry of Emergency Situation (MES) report—Monitoring, forecasting hazardous processes and phenomena in the Kyrgyz Republic, Book of the ministry of emergencies, 2014 г., UDK 502/504, BBK 20.1
4. Morris BL, Darling WG, Gooddy DC, Litvak RG, Neumann I, Nemaltseva EJ, Poddubnaia I (Jan 2006) Assessing the extent of induced leakage to an urban aquifer using environmental tracers: an example from Bishkek. Hydrogeol J. https://doi.org/10.1007/s10040-005-0441
5. NEHRP (National Earthquake Hazards Reduction Program) (2003) Recommended Seismic Provisions for New Buildings and Other Structures (FEMA P-750) 2003 Edition. BSSC. Washington. 338 c
6. Okada H (2003) The microtremor survey method, geophysical monograph series number 12, Society of Exploration Geophysicists
7. Okada H (2006) Theory of efficient array observations of microtremors with special reference to the SPAC method. Explor Geophysicists 37:73–85
8. Orunbaev S (2018) Improvement of methods on seismic hazard assessment with example of a certain local sites of Kyrgyzstan, Ph.D. thesis, Institute Dynamics of Geosphere, Russian Academy of Sciences, Moscow, p 146
9. Parolai S, Orunbaev S, Bindi D, Strollo A, Usupaev Sh, Picozzi M, Di Giacomo D, Augliera P, D'Alema E, Milkereit C, Moldobekov B, Zschau J (2010) Site effect assessment in Bishkek (Kyrgyzstan) using earthquake and noise recording data. Bull Seismol Soc Am. Submitted to BSSA, pp 3068–3082
10. Pozdnyakova L, Pozdnyakov A, Zhang R (2001) Application of geophysical methods to evaluate hydrology and soil properties in urban areas. Urban Water 3:205–216
11. Pozdnyakov AI, Pozdnyakova LA, Karpachevskii LO (2006) Relationship between water tension and electrical resistivity in soil. Eurasian Soil Sci 39(Suppl 1):78–83
12. Pozdnyakov AI, Gulalyev Ch.G (2008) Electrophysical properties of some soils. M.-Bacu: OOO. In: Smernikova SA, Pozdnyakova AI, Shain EV (eds) Assessment of urban soil

flooding by electrophysical methods. Soil Science, on the occasion of its 100th anniversary of Vladychenskogo S.A., №10, c.1–8
13. Special report of the Ombudsman (Akyikatchy) of the Kyrgyz Republic: "New buildings in Bishkek: the right to adequate housing and other social, economic and cultural rights". Bishkek, 2017 г. « Adilogly » , 2004, 239 c
14. Yokoi T, Margaryan S (2008) Consistency the Spatial Autocorrelation method with Seismic Interferometry and its Consequence. Geophys Prospect 56:435–451

Part IV
Ecological Baseflow/Environmental Flow to Keep the Health of Rivers and Lakes

Chapter 11
Environmental and Water Resources Management Study of Ghorband-Panjshir and Kunduz Rivers Basins, Afghanistan

H. Habib, Akhundzada Noor Ahmad, W. Wafa, and M. H. Hairan

Abstract In this study, Integrated Water Resources Management (IWRM) and Environmental Assessment of Kundoz and Panjshir-Ghorband river basins are conducted. The Kunduz river basin is a tributary of the Amu-Darya in North Afghanistan, and the Panjshir-Ghorband River is a tributary of the Kabul river basin. In this study, for land cover, climate change impacts, water resources, and hydrological investigations purposes, long-term hydrometeorological data were analyzed. Water quality and River discharge were measured in the field, and a socioeconomic study was conducted in the field through questionnaire use supplemented by background data from Central Statistics Organization (Central Statistical Organization (CSO) Afghanistan, Statistical Yearbook 2015–2016; 2016.). Investigations results show that, since the 1960s, mean annual temperature gradually increased and, inversely, precipitation and river discharge gradually decreased. The environmental degradation and resultant climatic and natural disasters have deeply affected the livelihood and wellbeing of the communities within these river basins with regards to water, agriculture, and ecosystem life. In conclusion, this study proposes steps for IWRM and climate and disaster mitigation including climate adaptation through water resources conservation and environmental management.

Keywords IWRM · Land cover · Climate change impacts · Socioeconomic investigation

11.1 Introduction

In Afghanistan, the Hindukush Mountain Range is the continuation of the Himalayan orogeny that happened North-East to South-West of the country. It gives Afghanistan harsh and steep mountain topography and divides Afghanistan's surface water into five major river basins. These are the Kabul, Helmand, Harirud-Murghab, Northern and Amu-Darya river basins. Except for the northern river basin, the remaining four major rivers leave the national borders and flow into the neighboring countries the

H. Habib (✉) · A. Noor Ahmad · W. Wafa · M. H. Hairan
Kabul University, Kart-e-Char, Kabul, Afghanistan

Z. E. Kulenbekov and B. D. Asanov (eds.), *Water Resource Management in Central Asia and Afghanistan*, Springer Water, https://doi.org/10.1007/978-3-030-68337-5_11

downstream. The Amu-Darya and Kabul rivers Basins are the main water resources used for drinking, irrigation, hydropower and environmental purposes in Afghanistan and down River. Approximately, 80% of the population in Afghanistan depends on agriculture for their livelihoods, and agriculture contributes to almost half of the GDP [1]. Recent climate change impacts have changed the hydrological conditions and land cover of the main river basins [2]. Glaciers and permafrost in the Hindukush mountains have melted [3]. The decrease in precipitation and increase in glaciers melt has decreased water flow in the Amu-Darya and Kabul river basins. Alongside this, environmental degradation, erosion and drought have severely affected the watershed of both rivers basins [4].

While water resource conservation, management, and sustainable development are an integral part of economic development and environmental restoration in Afghanistan, there has been no comprehensive study conducted on climate change impacts, IWRM, land cover, and environmental degradation [5]. The resultant lack of climate change impact, hydrology, and water quality data is one of the main challenges for sustainable water resources management. Most hydrological and climate data collection in Afghanistan was interrupted in the early 1980s as a consequence of war, only partially resuming after 2004 [6]. The purpose of the study is to conduct Integrated Water Resources Management (IWRM) and Environmental Assessment of the Kunduz and Panjshir-Ghorband river basins. The Kunduz river basin is a tributary of the Amu-Darya in North Afghanistan and the Panjshir-Ghorband river basin a tributary of the Kabul River basin.

11.2 The Area of Investigation

11.2.1 Topography and Climatic Condition

Figure 11.1 shows the hydrology and geographic location of the study area. Both the Panjshir-Ghorband and Kunduz subriver basins are originated from the Hindukush Mountains. The Panjshir-Ghorband river originates from the south and the Kunduz river from the north side of the Hindukush Mountains and flow through northern Afghanistan, finally reaching the Amu-Darya. The Panjshir-Ghorband watershed encompasses about 12,963 km^2, and the Kunduz watershed covers 28,023 km^2 in area [7].

As the relief of the river basins consists of high-altitude mountainous areas ending with lowland areas, precipitation and temperature have high variation within the basins. The annual average precipitation and temperature in the Jabalusarag area, located in the middle of the Panjshir-Ghorband river basin, is 470 mm and 14.4 °C for 2018. The average annual precipitation and temperature in the Kunduz area is 325 mm and 16.8 °C, respectively [8].

Both of the river basins consist of a number of tributaries. The hydrology of the basins is mainly controlled by the high mountains of the Hindukush [7]. Upstream,

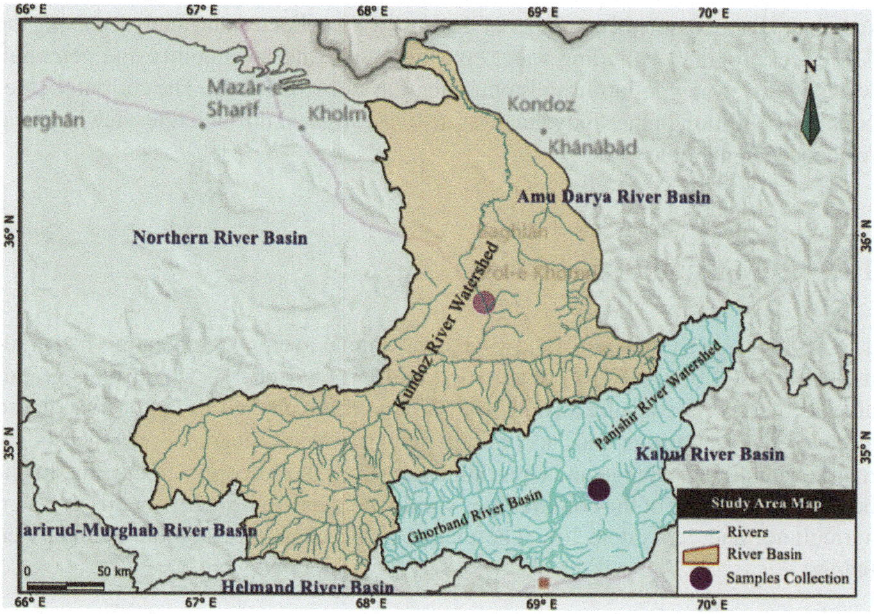

Fig. 11.1 Ghorband-Panjshir and Kunduz river Watershed locations made by Noor Ahmad

channels are generally narrow and flow throughout the whole year. Runoff regimes are largely controlled by snow-melt and high discharge happens from April to June [9]. Precipitation in these Basins mainly occurs in the form of rain, drizzle, snowfall, and hail, and is high during the winter months. The area near the Rivers and the floodplains consists of highly fertile soils with productive agricultural land which also comprises the main economic center of the basins. The higher altitude areas in the Basins are partly used for rain-fed agriculture, but mostly consist of deforested land [1].

11.3 Methodology

For the study, an Electromagnetic Velocity Meter, Cyber Scan 600 Series, and portable Waterproof Handheld were used to determine river discharge and physical parameters in the field. For climate change impacts assessment, existing historical hydrometeorological data, such as air temperature, precipitation, evaporation, river discharge, water quality, humidity, and air pressure were analyzed and interpreted. Water chemical analysis was conducted in the environmental laboratory of the American University of Central Asia in Bishkek, Kyrgyzstan. For Landcover investigation, remote-sensing data and satellite imagery from NASA, USGS, UNOSAT, MODIS were analyzed using Manifold GIS and ArcGIS packages. For environmental and social impact assessment, a number of datasets were collected and analyzed using the MS Excel program, including water consumption, water availability and potential, water quality, water pollution and contamination and census data. The current socioeconomic condition of the study area was also investigated through interviewing local residences in the field.

11.4 Results and Discussions

The results showed that, since the 1960s, the annual average temperature has gradually increased inversely to precipitation and River discharge which have gradually decreased. The results also showed a large scale change in land cover due to climate change impacts and, accordingly, environmental degradation. The above-mentioned climate change and environmental issues have deeply affected the livelihood and wellbeing of the communities within the river basins with regards to water, agriculture, and ecosystem life. The main climate change results are summarized below.

11.4.1 Climate Change Impacts

Figures 11.2 and 11.5 show a gradual increase in mean annual temperature in both river basins. The data was recorded at Jabalusarage and North Salang meteorological stations. The Hydrometeorological data is received from the Ministry of Energy and Water of Afghanistan [10]. Figures 11.3 and 11.6 show a gradual decrease in the long-term precipitation in the basins. The break in the graph indicates a data gap from about 1980 to 2000 that is due to war and conflict in the country. Figures 11.4 and 11.7 show a long-term gradual decrease in the discharge of both rivers. Kulukh

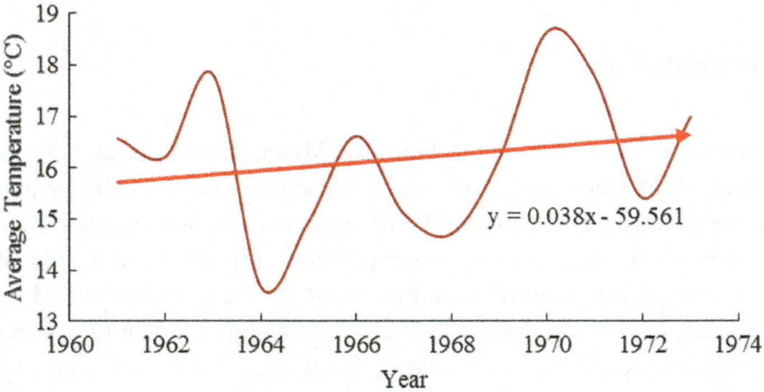

Fig. 11.2 Mean annual temperature at Jabalusarage Station showing a gradual increase in temperature since 1960

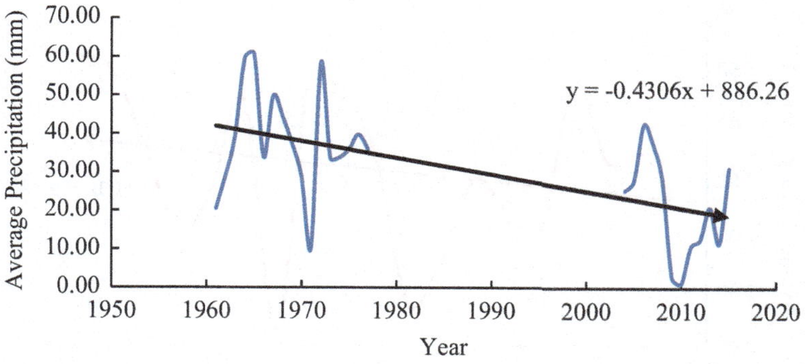

Fig. 11.3 Mean annual precipitation (mm) at Jabalusarage Station showing a decrease in precipitation since 1950

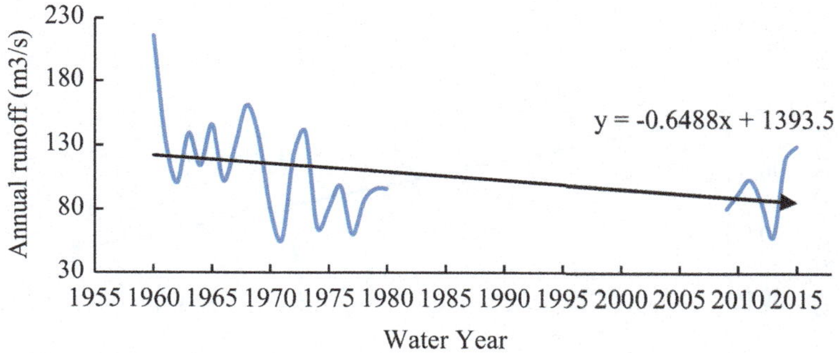

Fig. 11.4 Long-term mean annual runoff at the Naghlu gauging station showing a gradual decrease in river runoff since 1955

Tepa gauging station is located at the end of the Kunduz river in the Amu-Darya Basin and the Naghlo station is located at the end of the Panjshir-Ghorband river at the Kabul River. The figures thus show the final discharge of the river basins. Results from the other stations in the basins are the same.

11.4.2 Socioeconomic Status

The Kabul river basin population has been estimated by the CSO Afghanistan (2016) in 2015. The total population of 9.2 mln people is directly affected by the Kabul river. The Baghlan provincial capital is Pul-e-Khumri and has about 79 883 inhabitants. A typical household in Panjsher has approximately 6.6 persons, which is slightly higher than the national average of 6.3. According to the CSO Afghansitan (2016), Baghlan has a total population of 741,690 [11].

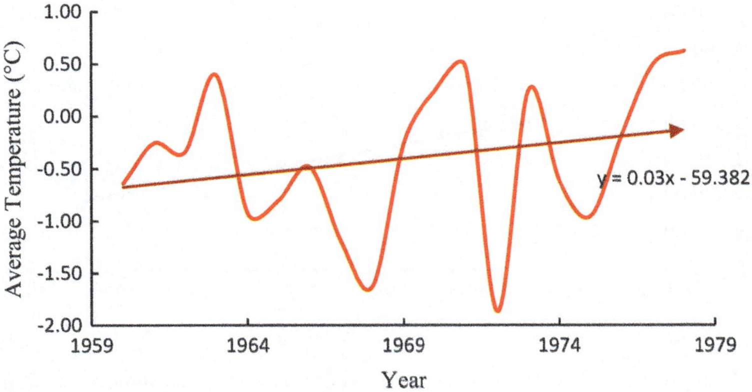

Fig. 11.5 Mean annual temperature at North Salang Station showing a gradual increase since 1960

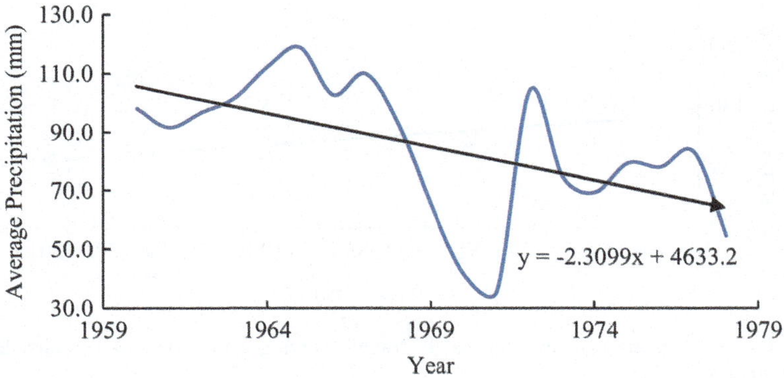

Fig. 11.6 Mean annual precipitation at North Salang Station showing decrease since 1959

In this study, socioeconomic assessment is estimated via social and economic indicators, such as livelihood, food security, and agriculture. In general, in both river basins, most livelihoods rely on agriculture. To some extent the living conditions in each river basin are different. For example, life conditions in the Pul-e-Khumri area are better than in the Sayad. Besides the agricultural activities in Sayad there are some tourist and recreational activities. In Pul-e-Khumri, the capital of Baghlan province, there are business activities, which is a clear indication of good economic condition. Therefore, climate change affects existing food security and impacts heavily those dependent on the agricultural economy within the Basins. The distributional effects are more likely to fall upon women and children, and upon those involved in subsistence agriculture or pastoralism. The effects of environmental degradation and lower agricultural output reduce the availability of animal feed, and increase the effort required for livestock husbandry which directly affects the livelihood of the people in the Basin.

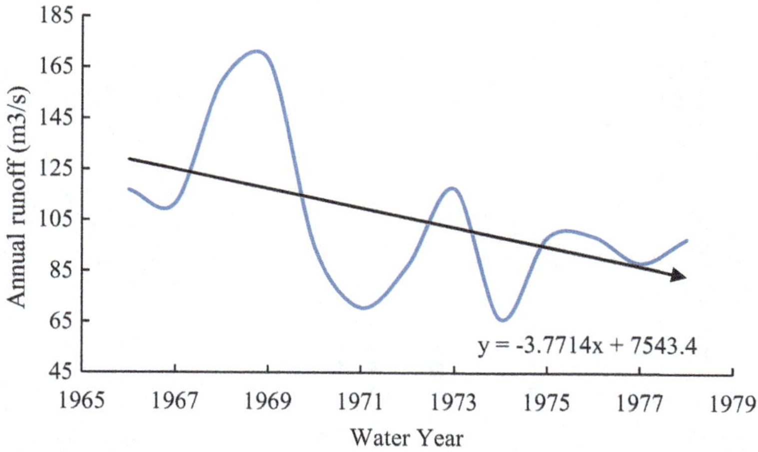

Fig. 11.7 Long-term mean annual runoff at the Kulukh Tepa gauging station showing a gradual in River runoff since 1965

11.4.3 Water Quality

Water quality was measured in the field and laboratory for both river basins. Almost all the physical parameters shown in Table 11.1 were measured in the field with the exception of the chemical analysis conducted at the American University of Central Asia in Bishkek, Kyrgyzstan. As shown in Table 11.1, all the parameters are within the norms and well below the standards.

11.5 Conclusion

Climate change impacts affect the existing food security and impact heavily those dependent on the agricultural economy. The effects of climate change induce frequent and intense droughts in the river basins. Precipitation has decreased, resulting in water resources depletion, and threatened groundwater supply for entire communities in the basins. The climate change impacts and environmental degradation from war and conflict in the country have deeply affected vegetation cover, agriculture, and forestry in the river basins, which has affected climate change adaptive capacity. Rising winter and spring temperatures lead to rapid and earlier snowmelt, creating risks of flash flooding. The impact of increasingly frequent flash floods is exacerbated by drought, which has the effect of hardening soils and reducing their permeability. All these impacts lead to a range of humanitarian crises and socioeconomic problems including disease, population displacement, lost livelihood, and conflict.

This study proposes steps for IWRM of both river basins. There is good potential for climate change mitigation and adaptation and natural disaster management

Table 11.1 The physics–chemical properties of Panjshir and Kunduz rivers

No	Parameters	Norms	Samples result		
			Panjshir river, Sayad	Kunduz river, (upper)	Kabul river, Sarobi
Physical parameters					
1	pH	Within 6–9	8.25	8	7.15
2	TSS mg/l	–	< 1	< 1	< 1
3	TDS mg/l	–	298	405	521
4	Temperature (°C)	5 to 15	16	17	
5	Conductivity (µs/cm)	1500	384.3	440.9	
6	NaCl (mg/l)	–	358.8	416.4	
7	Resistivity (KΩ)	–	1.357	1.184	
8	DO mg/l	–	8.95	5.11	
Toxic metals					
9	Molybdenum (Mo) µg / l	70	0.30	0.97	9.80
10	Cadmium (Cd), µg / l	1	0.07	0.06	0.01
11	Copper (Cu), µg / l	1000	0.55	3.39	3.31
12	Nickel (Ni), µg / l	20	2.52	0.40	0.34
13	Lead (Pb), µg / l	10	0.92	0.65	0.75
14	Chromium (Cr), µg / l	10	1.57	2.20	1.27
15	Zinc (Zn), µg / l	5000	3.92	9.35	8.62

through ecosystem development within the existing water resources conservation and environmental management. This requires intensive research work and community-based participation approaches. Some steps are proposed as recommendations for filling the gaps in this study and conducting future scientific research. These are:

- Existing field investigation equipment must be well installed and calibrated and data will be recorded regularly and precisely.
- The correlation between historically recorded data and online recommended data needs more research and investigation as they have not matched.
- To acquire accurate data and address current gaps, we must conduct investigations and receive accurate estimated data from the relevant international agencies.

Acknowledgements This research was accomplished thanks to the financial support of USAID and PEER regional project.

References

1. World Bank (2004) Water resource development in Northern Afghanistan and its implications for Amu Darya Basin, World Bank Working Paper No. 36. Washington, D.C. 20433, U.S.A
2. Afghanistan Intended Nationally Determined Contribution (INDC) (2015) Submitted to the United Nations framework convention on climate change (UNFCCC)
3. Aich V, Akhundzadah NA, Knuerr A, Khoshbeen AJ, Hattermann F, Paeth H, Scanlon A, Paton EN (2017) Climate change in Afghanistan deduced from reanalysis and coordinated regional climate downscaling experiment (CORDEX) South Asia simulations. J Clim 5(35). Online (www.mdpi.com/journal/climate)
4. Environment and Security in the Amu Darya Basin (2011) Report prepared in corporation with UNEP, UNDP, UNECE, OSCE, REC, NATO
5. Habib H, Anceno AJ, Fiddes J, Beekma J, Ilyuschenko M, Nitivattananon V, Shipin OV (2013) Jumpstarting post-conflict strategic water resources protection from a changing global perspective: Gaps and prospects in Afghanistan. J Environ Manage 129:244–259. https://doi.org/10.1016/j.jenvman.2013.07.019
6. Mack TJ, Akbari MA, Ashoor MH, Chornack MP, Coplen TB, Emerson DG, Hubbard BE, Litke DW, Michel RL, Plummer LN, Rezai MT, Senay GB, Verdin JP, Verstraeten IM (2010) Conceptual model of water resources in the Kabul Basin. U.S. Geological Survey Scientific Investigations Report, Afghanistan, pp 2009–5262
7. Favre R, Kamal GM (2004) Watershed atlas of Afghanistan. Helolse condroyer, atlas communication 2004. FAO (1993) The Islamic Republic of Afghanistan Watershed Atlas, Food and Agriculture Organization of the United Nations (FAO) Rome, Italy
8. "https://www.weather-atlas.com/en/afghanistan/"
9. Lashkaripour GR, Hussaini SA (2007) Water resource management in Kabul River basin, eastern Afghanistan. Environmentalist 28(3):253–260. https://doi.org/10.1007/s10669-007-9136-2
10. Ministry of Energy and Water (2016) Hydrological Year Book. Islamic Republic of Afghanistan, Kabul
11. Central Statistical Organization (CSO) (2016) Afghanistan, Statistical Yearbook 2015–2016

Chapter 12
Physical–Chemical Properties of the Kyzyl-Suu River and Its Tributaries, Kyrgyzstan

Zheenbek E. Kulenbekov and Baktyiar D. Asanov

Abstract The article presents the results of studies of the Kyzyl-Suu River and its two tributaries, the Daara River (Daroot-Korgon village) and the Aryk-Suu River (Sary-Tash village). The study was carried out as part of the PEER NAS USAID project: "Integrated Water Resources Management of the Amu Darya and Kabul Rivers". Physical parameters of water, such as pH, temperature, electrical conductivity, resistance, etc., obtained in the field using the CyberScan Eutech PCD 650 instrument recommended by the US Environmental Protection Agency. Anions were measured in the chemical laboratory of the American University of Central Asia (AUCA). Bioindication of hydrobionts was carried out in the laboratory of ichthyology and hydrobiology of the Biological and Soil Institute of the NAS of the Kyrgyz Republic. The measurement was carried out at two points in the Kyzyl-Suu River (near the villages of Sary-Tash and Daroot-Korgon) and at two points on the Daara River (before and after the village of Daroot-Korgon), respectively, two points on the Aryk-Suu River (before and after village of Sary-Tash). Water samples were also taken. The results of the analysis for anions show that the content of chlorides, sulfates, nitrates, bicarbonates and phosphates, according to SanPin 2.1.4.559-96, GN 2.1.2.689-98, does not exceed the maximum permissible concentration (MPC) at any of the water sampling points. According to the results of bioindication, we can conclude that the water from the rivers is "clean". Abnormal values of pH, temperature, electrical conductivity, resistance, total dissolved salts, NaCl, and dissolved oxygen were not observed at the measurement points. However, it should be noted that all the measurement results have a tendency to increase after each settlement, albeit slightly, i.e., there is a certain anthropogenic impact on these rivers.

Keywords Anions · AUCA · CyberScan eutech PCD 650 · PEER NAS USAID

Z. E. Kulenbekov (✉) · B. D. Asanov
American University of Central Asia, 7/6 Aaly Tokombaev Str, Bishkek, Kyrgyzstan
e-mail: kulenbekov_z@auca.kg

Z. E. Kulenbekov and B. D. Asanov (eds.), *Water Resource Management in Central Asia and Afghanistan*, Springer Water, https://doi.org/10.1007/978-3-030-68337-5_12

12.1 Introduction

The implementation of the PEER NAS USAID project, "Integrated Water Resources Management and Strategic Environmental Assessment of the Amu Darya and Kabul Rivers," began in 2017 and will continue until the end of 2019. Three universities are participating in the project: Kabul University (Afghanistan); Mining and Metallurgical Institute of Tajikistan (Tajikistan); American University of Central Asia (AUCA) (Kyrgyzstan). AUCA is the regional coordinator of the project.

Measurements of physicochemical parameters and bioindication of the Kyzyl-Suu River and tributaries of the Daara and Aryk-Suu were not carried out. On the Kyzyl-Suu River, only water flow is measured. In the Kyrgyz Republic, research under the project is carried out in the Chon-Alai region, the village of Daroot-Korgon and in the Alai region, the village of Sary-Tash. These villages are located in the immediate vicinity of the Kyzyl-Suu River. The Kyzyl-Suu River is formed within the Alai Valley. The average water flow is quite high, about 65 m^3/s, and it eventually flows into the Vakhsh River outside of Kyrgyzstan's borders. The Vakhsh River is the right tributary of the Amu Darya River.

This article presents the results of bioindication and some physicochemical analyses of the water of the Kyzyl-Suu River and its two tributaries, the Daara and Aryk-Suu Rivers.

The purpose of this article is to study the physicochemical characteristics of the Kyzyl-Suu River and its tributaries, to evaluate and analyze the main changes in the physicochemical composition of the water of rivers and its tributaries, as well as its current state. Based on the results of a literature review of this kind, an assessment and analysis of the Kyzyl-Suu River and its tributaries were not found, and therefore, this study may be the first of its kind carried out in this region of Kyrgyzstan.

12.1.1 Research Methodology and a Brief Description of the Determined Parameters

For sampling and measurement, two points were identified in the Kyzyl-Suu River, two sampling points on the Daara River and, accordingly, two points on the Aryk-Suu River, all points are shown in Fig. 12.1.

1. Kyzyl-Suu 1—located on the Kyzyl-Suu River, after the settlement of Sary-Tash;

   1.1 Aryk-Suu 1—located on the Aryk-Suu River, to the village of Sary-Tash;
   1.2 Aryk-Suu 2—located on the Aryk-Suu River, after the settlement of Sary-Tash;

2. Kyzyl-Suu 2—located on the Kyzyl-Suu River, after the village of Daroot-Korgon;

   2.1 Daara 1—located on the Daara River, to the village of Daroot-Korgon;

Fig. 12.1 Measurement and sampling points

2.2 Daara 2—is located on the Daara River, after the village of Daroot-Korgon.

Using the modern instrument, CyberScan Eutech PCD 650, data on pH, electrical conductivity, resistance, dissolved salts, oxygen, and water temperature was obtained at the sampling site. The CyberScan Eutech PCD 650 is very user-friendly and has a wide measuring range [1].

The pH value characterizes the acid-base balance in water and it is one of the most important indicators of water quality. The development of aquatic organisms, the formation of various elements and much more depends on pH. Generally, the pH values of natural water bodies range from 6.3 to 8.5. The pH value is affected by the state of carbonate equilibrium, the intensity of photosynthesis, and the decomposition of certain substances, as well as the content of humic substances [2].

Electrical conductivity is the quantitative characteristic measuring water's ability to conduct electric current. In most cases, the electrical conductivity of land surface water is an approximate measurement of the concentration of inorganic electrolytes in water, namely, cations and anions. The conductivity of surface water of the land depends mainly on the water's mineralization and usually ranges from 50 to 10,000 μS/cm [3]. Naturally, water temperature and dissolved oxygen are of great importance for aquatic organisms. Water temperature is indicated in degrees Celsius (°C), and dissolved oxygen in percent (%) and milligram per liter (mg/L).

Sampling for the analysis was carried out according to the rules of sampling, storage, and transportation of samples was conducted in accordance with GOST R

51592, GOST 17.1.5.05, GOST 24902, ISO 5667. Water samples were analyzed for chlorides, sulfates, nitrates, etc., at the AUCA chemical laboratory. All parameters selected for analysis are arguably the most important indicators of water quality. Chlorides and sulfates in drinking water do not have a toxic effect on humans, but they worsen the taste of water, the taste of sulfates and chlorides arises at a concentration of 250–400 mg/L. The increased nitrate content in water can serve as an indicator for the pollution of a reservoir as a result of the spread of fecal or chemical pollution (agricultural or industrial). The high nitrate content in water stimulates the mass development of aquatic vegetation. Drinking water and foods containing an increased amount of nitrates can cause various diseases in humans [4].

In addition to chemical assessment, biological methods for determining quality play an important role in determining water quality. Assessing the quality of the habitat and its individual characteristics according to the state of biota in natural conditions is called bioindication. Bioindicators are defined as species or groups of species or communities by presence, degree of development, and changes in morphological, structural, functional, and genetic characteristics. These aforementioned bioindicators help to judge the quality of water and the state of an ecosystem [5].

The selection of hydrobionts was carried out according to the directions of ISO 7828. Biondication was carried out in the laboratory of ichthyology and hydrobiology at the Biological Soil Institute in the NAS of the Kyrgyz Republic.

12.2 Results

To clarify, all data obtained in the field are averaged and shown in Table 12.1. As shown here by the results of measurements, no anomalous phenomena were revealed in the sampled river waters. In all sampled rivers, the average water temperature rises downstream. This rise in temperature can be explained by the change in altitude changing air temperatures. According to temperature data from the sampled rivers, there is no reason to discuss temperature pollution.

In accordance with the requirements for the composition and properties of water and water within recreation, as well as for other water bodies, the pH should not go beyond the range of values 6.5–8.5 [6]. As shown, the results are not outside of the normal range of pH values. Water at pH 6.5–7.5 is neutral and is explained by the presence of $Ca(HCO_3)_2$, $Mg(HCO_3)_2$ in them.

The electrical conductivity of natural water depends mainly on the concentration of dissolved mineral salts and temperature. Natural waters are mainly solutions of mixtures of strong electrolytes. The mineral part of the water is made up of Na + , K + , Ca_2 + , Cl−, SO_4^{2-}, HCO_3− ions [7]. As can be seen from Fig. 12.2, the electrical conductivity of downstream increases, which indicates an increase in the salinity of water.

In natural water, the value of the redox potential ranges from minus 400 to plus 700 mV. Redox potential is determined by the totality of the oxidation and reduction

Table 12.1 Average values for sampling points measured with the CyberScan Eutech PCD 650

Parameters	Kyzyl-Suu 1 (Sary-Tash)	Kyzyl-Suu 2 (Daraot-Korgon)	Aryk-Suu 1	Aryk-Suu 2	Daara 1	Daara 2
pH	7.20	7.63	7.15	7.67	7.18	7.22
Temperature (°C)	10.90	11.27	6.20	6.53	9.23	12.57
* RP, mV	−10,45	−37,17	−8,80	−39,80	−8,13	−12,77
Conductivity (μS)	680,10	945.83	317,93	429,35	390,43	386,30
**TDS (mg/L)	372,32	472,91	400,37	412,70	374,17	370,47
NaCL (mg/L)	352,83	410,93	373,90	395,85	352,00	353,67
***DO (%)	70,37	74,03	69,53	77,05	68,40	72,33
***DO (mg/L)	5,44	5,68	5,53	5,66	5,37	5,51

* RP—redox potential
** TDS—Total Dissolved Solids
*** DO—Dissolved Oxygen

processes occurring in it. Under equilibrium conditions, the redox potential characterizes the medium immediately relative to all elements with variable valency. The study of redox potential makes it possible to identify natural environments where the existence of chemical elements with variable valency in a certain form is possible, as well as to highlight the conditions under which metal migration is possible [8]. According to the types of geochemical conditions in natural waters, the RP value is characterized by values <0 refers to the reductive type. The water at all measurement points has been designated as the reducing type, since it has negative values.

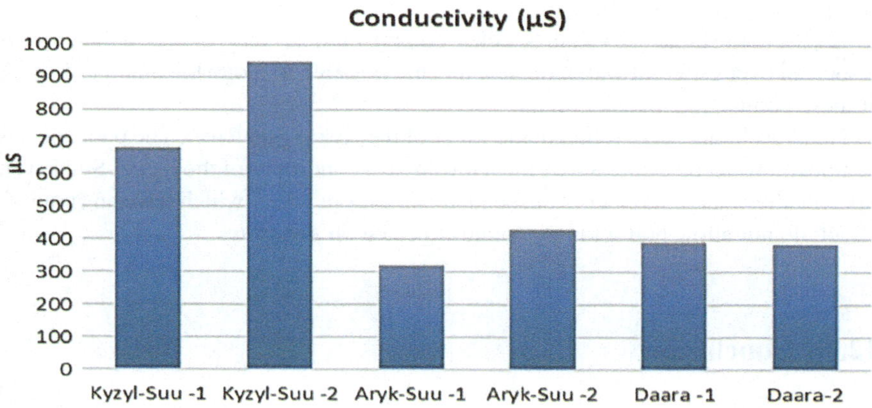

Fig. 12.2 Electrical Conductivity of rivers

Table 12.2 The results of chemical analyzes of water

Parameters, mg/L	Kyzyl-Suu 1	Kyzyl-Suu 2	Aryk-Suu 1	Aryk-Suu 2	Daara 1	Daara 2	MPC, mg/L
Chlorides (Cl^-)	133,13	266,25	80	53,25	44,38	85,2	350
Sulphates (SO_4^{2-})	<100	<100	<100	<100	<100	<100	500
Nitrates (NO_3^-)	1	1	1	1	1	3	45
Ammonium (NH_4^+)	0,7	1	0,2	0,5	1,5	1,5	2,5
Hydrocarbonates (HCO_3^-)	244	366	183	213,5	152,5	152,5	1000
Orthophosphates (PO_4^{3-})	0,2	0,2	0,2	0,2	0,2	0,5	3,5

The average values of chlorides, sulfates, nitrates, and ammonium are listed in Table 12.2. According to these parameters, we can confidently say that at no point the MPC exceeded. The value of chlorides almost doubles in the Daara River, after the village of Daraot-Korgon. We observe the same trend downstream in the Kyzyl-Suu River. But in the Aryk-Suu River, a decrease in chlorides is observed downstream.

The content of sulfates and nitrates at all points of sampling remains unchanged. An exception is the nitrate content on the Daara River, which doubles after the village of Daroot-Korgon. The ammonium content at almost all points of sampling increases downstream, although not significantly. The content of orthophosphates and bicarbonates also slightly increases downstream and does not exceed the norm at any point.

As seen in Table 12.3, the structure of zoobenthos is extremely mosaic (i.e. at various sampling stations, various species of benthic organisms are present). In total, 18 species of invertebrates were found in the Daara River. The water quality in general, according to the table, can be assessed as "pure", according to the generally accepted classifications [9].

As seen in Table 12.4, the species composition is very poor, zoobenthos is dispersed over different watercourses. It is not possible to judge the quality of water at these points, due to lack of data.

No hydrobionts were found in the water of the Kyzyl-Suu River. The poor species composition can be explained by the hydrological conditions of the Kyzyl-Suu River. Specifically, a large number of suspended particles in water, which have an "emery" effect, do not allow bottom invertebrates to catch on substrates.

12.3 Conclusion

The results of the analyses of anions show that at no water sampling point do the contents of chlorides, sulfates, nitrates, bicarbonates, and phosphates according to SanPin 2.1.4.559–96, GN 2.1.2.689–98 exceed MPC [10]. Based on the results of bioindication, we can conclude that the water in the Daara River is "clean". It was

Table 12.3 The structure of bottom invertebrates (zoobenthos) of the Daara River

Species composition	Daara 1	Daara 1 (silt)	Daara 1 (flushing from stones)	Daara 2	Daara 2 (flushing from stones)
Chironomid larvae:					
Diamesa insignipes	–	+	–	+	+
Diamesa pseudostylata	+	+	+	+	+
Orthocladius thienemanni	–	–	–	+	–
Eukiefferiella sp.	–	+	–	–	–
Thienemannimyia sp.	–	+	–	–	–
Simulid larvae:					
Metacnephia sp.	+	–	+	+	+
Simulium sp.	+	–	–	–	–
Larvae of other dipterans:					
Deuterophlebia mirabilis	+	–	–	–	–
Blepharicera asiatica	–	–	+	–	–
Athericidae basilica	+	–	–	–	–
Mayfly Larvae:					
Epeorus (Iron) из группы montanus	+	+	–	+	–
Epeorus (Ironopsis) rheophilus	–	–	–	–	+
Rhithrogena tianshanica	–	–	+	–	–
Ecdyonurus sp.	–	–	–	+	–
Baetis venustulus	+	–	–	–	–
Baetis kulindrophtalmus	+	+	–	–	–
Springtime larvae:					
Filchneria mongolica	–	–	–	–	+
Amphinemura sp.	–	–	–	+	–
Total species: 18	8	6	4	7	5

Table 12.4 The structure of bottom invertebrates (zoobenthos) of the Kyzyl-Suu and Aryk-Suu Rivers

Species composition	Aryk-Suu 1 (flushing from stones)	Aryk-Suu 1	Aryk-Suu 2 (flushing from stones)	Aryk-Suu 2
Chironomid larvae:				
Diamesa pseudostylata	–	–	–	+
Syndiamesa branickii	–	+	–	–
Eukiefferiella alpestris	–	–	+	–
Larvae of other dipterans:				
Tipula (Yamatotipula) sp.	–	–	–	+
Mayfly Larvae:				
Epeorus (Iron) из группы montanus	+	–	–	–
Baetis kulindrophtalmus	–	–	+	–
Caddis flies:				
Hydropsyche guttata	+	–	–	–
Integripalpia[a]	–	–	+	–
Total species: 8	2	1	3	2

[a] the larvae were damaged, so it was not possible to determine either the genus or species

not possible to determine the water quality in the Kyzyl-Suu River using bottom invertebrates, since no zoobenthos were found in the river. In the Aryk-Suu River, it is also not possible to judge the quality of water due to the poor species composition found there. Abnormal values of pH, temperature, electrical conductivity, resistance, total dissolved salts, NaCl, and dissolved oxygen were not observed at the measurement points. However, it is worth noting that all values, albeit not significantly, tend to increase after each settlement. From which we can conclude that there is a slight anthropogenic impact on the rivers. This is most likely due to animal husbandry, since there are no industrial emissions in the settlements near the sampled rivers. This study indicates that there is a need for annual monitoring of the water quality of the Kyzyl-Suu River and its tributaries.

References

1. http://mankor.ua/catalog/yams_multimeters/yame_handheld_multimeters_eutech/2019/11/08
2. RD 52.24.495-2005 "Hydrogen indicator and electrical conductivity of water. Measurement method by electrometric method." (approved by the Russian hydromet 15.06.2005)

3. Muravyov AG (2011) Guidelines for water analysis. Drinking and natural water, soil extracts. "Crismas +", St. Petersburg
4. Field eco-school. LLC Ilim NGO, PA Akmena, OSCE Center in Bishkek
5. Guardian MG (2004) Bioindication of pollution. Publishing House of St. Petersburg State University, SPb
6. Kosov VI, Ivanov VN (1995) Protection and rational use of water resources. Part 1 Surface water protection: study. Allowance.- Tver. State Tech. University
7. Semenov AD (1977) Manual for the chemical analysis of surface water of land. Hydrometeoizdat
8. Nikanorova AM (1988) Handbook of hydrochemistry. Hydrometeoizdat
9. Kustareva LA (1997) Lemzina LV. Life in the waters of Kyrgyzstan. Ilim, Bishkek
10. SanPin 2.1.4.559-96, GN 2.1.2.689-98

Chapter 13
Water Reuse for Irrigation in Rural Areas in Japan Experience from Japan to Central Asia

Koji Hamada

Abstract As distinct from Central Asia, Japan has three types of wastewater treatment systems: municipal wastewater treatment, rural sewage treatment, and domestic wastewater treatment systems. Municipal wastewater systems are large-scale systems and the treatment plants are constructed at the downstream end of urban areas. Farmlands are usually far from treatment plants. In contrast, middle-scale rural sewage systems were installed in rural areas with the purpose of conserving the quality of irrigation water for farmland located around these treatment plants. The results of our test cultivation, the water quality of treated rural sewage effluent (TRSE) is suitable for agricultural reuse, and its negative effect on soil environments would be insignificant if any. Further, it is even suitable for reuse in greenhouses in Central Asian region. Positive reuse of TRSE, provided it is supplied close to farmlands, will secure a comparatively stable and safe water resource. Utilizing TRSE for irrigation purposes has the potential to be one of the solutions for water-stress improvement and aquatic environmental conservation around rural areas.

Keywords Rural sewage system · Municipal wastewater · TRSE · Greenhouse · Central Asia · Water recycle first section

13.1 Introduction

Japan, located in Monsoon Asia, receives relatively high rainfall by seasonal rain fronts and typhoons. Geographically the latitude location of Japan is the same as in Central Asia. Annual rainfall is high, around 1800 mm, but rainfall patterns have been changing in recent years. Rainfall patterns all over Japan indicate trends of increasing downpours and increasing 'no-precipitation' days. In this situation, efficient rainwater use is difficult, and restrictions on water intake from some rivers sometimes must be enacted. The Tone River, flowing near Tokyo, is an example of where these restrictions had to be enacted [1]. Changing rainfall patterns can

K. Hamada (✉)
Institute for Rural Engineering, NARO of Japan, Kan'nondai 2-1-6, Tsukuba Science City, Ibaraki 305-8609, Japan
e-mail: hamada34@affrc.go.jp

Z. E. Kulenbekov and B. D. Asanov (eds.), *Water Resource Management in Central Asia and Afghanistan*, Springer Water, https://doi.org/10.1007/978-3-030-68337-5_13

potentially cause water stress in a wider area beyond the generally recognized water-stressed regions. Under water-stressed conditions, restrictions on water intake can lead to a shortage of irrigation water, and there is an urgent need for alternative water resources for sustainable agriculture.

In recent years, efficient utilization of treated wastewater for irrigation has been globally focused on as a useful alternative water resource and many international guidelines regarding this have been published [2–4]. Japan has also been focusing on reusing treated wastewater as a useful alternative water resource in water-stressed regions such as remote islands. Positive reuse of treated wastewater should secure comparatively stable water resources, because water supply is of utmost priority. Furthermore, treated wastewater, which is usually nutrient rich, is one of the methods to bring nutrient loads to aquatic environments. If treated wastewater is utilized for irrigation, then the nutrients contained within it will be utilized for cultivation and should contribute to aquatic environmental conservation. Therefore, treated wastewater reuse for irrigation helps with both the alleviation of water stress and the conservation of aquatic environments.

13.2 Treated Municipal Wastewater Reuse

Japan has three types of wastewater treatment systems: municipal wastewater treatment, rural sewage treatment, and domestic wastewater treatment systems. Municipal wastewater systems are large-scale systems and the treatment plants are constructed at the downstream end of urban areas. Farmlands are usually far from treatment plants. In contrast, middle-scale rural sewage systems were installed in rural areas with the purpose of conserving the quality of irrigation water for farmland located around these treatment plants. Rural sewage systems are usually managed to only allow for the indirect usage of treated wastewater for irrigation. Domestic wastewater treatment systems are independent systems in individual houses simply to prevent water pollution by households.

Approximately 190 million tons per year of treated municipal wastewater, which was 1.4% of total treated municipal wastewater, was reused in 2008 [5]. According to the main applications of treated municipal wastewater, the utilization percentages were the following: 23% for landscaping, 23% for maintaining a minimum river flow, 23% for snow melting, and 9.4% for office use. Utilization for irrigation use accounted for only 1.4% of reused treated municipal wastewater.

Central Asia refers to the five former Soviet Union states Kazakhstan, Turkmenistan, Uzbekistan, Tajikistan, and Kyrgyzstan [6]. Central Asia stretches from the Caspian Sea in the west to China in the east and from Afghanistan in the south to Russia in the north.

Consequently, all five Central Asian states have experienced the crumbling of the Soviet Union in 1991, the following collapse of the economic system, and the subsequent socioeconomic upheaval [7].

The transition from a state planned to a market economy has meant changing patterns for basic services such as water supply and sanitation [8]. At present, Central Asia is home to a population of about 74 million (Kazakhstan 18.8 million, Kyrgyzstan 6.6 million, Tajikistan 9.6 million, Turkmenistan 5.8 million, and Uzbekistan 33.6 million) [9]. By 2040, the total population is expected to have increased to about 86 million. This together with countryside migration to urban areas will put enormous stress on water and infrastructure. Water supply availability in Central Asia is complicated due to that the region's two major rivers, the Syr Darya and Amu Darya, are transboundary. Thus, continual conflicts over the region's water resources have characterized the Central Asian states since the collapse of the Soviet Union [9].

13.3 Reuse of Treated Rural Sewage Effluent in Rural Areas

Rural sewage systems, which are middle scale and target to serve 20–1000 inhabitants, are installed in many rural areas in Japan. The annual amount of treated rural sewage volume is 350 million tons. One of the main purposes of rural sewage systems is the conservation of water environments in rural areas, including irrigation water quality, and 78% of rural areas use treated rural sewage effluent (TRSE) in agriculture [10]. TRSE flows into an irrigation canal and then the water is taken from the canal; this usage of TRSE is usually called 'indirect use for irrigation.' In this process, there is insufficient and ineffective utilization of nutrients of TRSE. Only the direct use of TRSE can lead to more effective utilization.

In Japan, application guidelines for irrigation water allow for 94% of water to be used for rice paddy cultivation and the other 6% for other crops. Crops other than rice are generally rain-fed due to relatively high precipitation, and the fact that irrigation equipment such as spray, sprinklers, and drip irrigation are not common in the fields. If we use TRSE for irrigation directly, greenhouse irrigation can become the major application. Valuable crops can be cultivated using high-cost TRSE.

Water resources in Central Asia are constituted by surface water and groundwater. The major part of potentially usable water comes from the large rivers of the region. All these rivers, however, are transboundary. The largest, the Amu Darya and the Syr Darya, flow through more than three countries.

The Amu Darya and Syr Darya account for 90% of Central Asia's river water. Especially, Turkmenistan and Uzbekistan are vulnerable due to that a major part of the water resources is generated outside of their respective territory. The Central Asian water problem is thus complex and also involving other transboundary stakeholders such as Russia, China, Afghanistan, Iran, and Pakistan. Other large rivers are the Irtysh and Ishym located in the eastern parts of Central Asia, the Chu and Talas located in the south, in the west the Ural can be found, and finally the Ishim and Tobol are found in the north [7]. The by far largest withdrawal of water for irrigation is done from the large rivers of the region. However, the area also contains countless smaller rivers and creeks that contribute to irrigated agriculture. The information on this, however, is often absent.

13.4 Reuse Test for Greenhouse Cultivation

Direct reuse of TRSE is effective in nutrient utilization even in greenhouses, but it could lead to another issue. TRSE has the possibility to impact soil environments in greenhouse irrigation. TRSE has higher ion concentrations, such as sodium and chloride, than water from natural resources. When TRSE is reused for open-field irrigation, the salts supplied by irrigation water to the farmland soil surface will be leeched-out by rainfall, and it is difficult for salt to accumulate in the soil. However, soil in greenhouses can have salt accumulation, as there is no rainfall in greenhouses. The effects of greenhouse irrigation on soil environments were examined by testing cultivation. We conducted two irrigation cases with TRSE and tap water for tomato cultivation and examined the effects of greenhouse irrigation on the soil environments (Fig. 13.1). We measured the sodium and chloride ion concentrations in soil extracts before and after cultivation.

Tomato cultivation:

- Irrigation by direct use of TRSE,
  (control: tap water)
- Drip irrigation in greenhouse,
- Irrigation control: pF value in the soil.
  (start point: pF 2.1).

The sodium and chloride ion concentrations were decreased by cultivation in both cases (Fig. 13.2). There were no significant changes in these concentrations between either cases. This result suggests that there is almost no negative effect on the soil environments, even in greenhouses, when using TRSE for irrigation through

Fig. 13.1 Test cultivation experiment, photo by Hamada

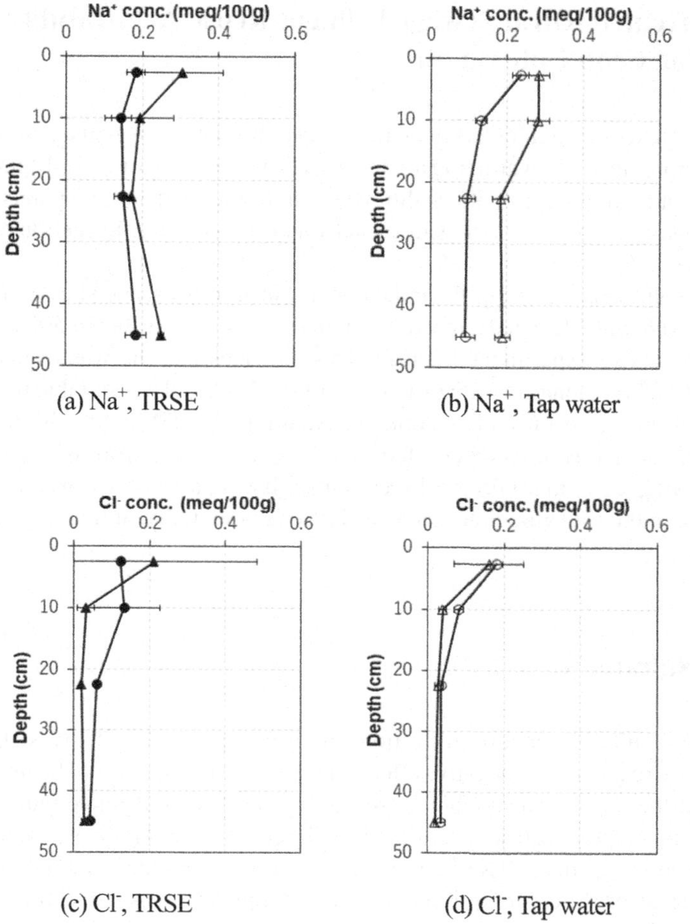

Fig. 13.2 Water-extracted ion concentrations from soils

the short term. The water quality of TRSE is suitable for agricultural reuse, and its negative effect on soil environments would be insignificant, if any. Further, it is even suitable for reuse in greenhouses. We evaluated only the short-term effects, but we require further evaluation on long-term irrigation. Positive reuse of TRSE, provided it is supplied close to farmlands, will secure a comparatively stable and safe water resource. Irrigation utilizing TRSE has the potential to be one of the solutions for water-stress improvement and the conservation of aquatic environments around rural areas in Japan.

13.5 Treated Rural Sewage Effluent Reuse Arbitrability for Crop Cultivation

The WHO and other guidelines indicate that health-related risk management is of the utmost importance when utilizing treated wastewater for irrigation, and pathogens are the main cause of concern with health risks. Rural sewage treatment plants generally employ a secondary treatment using an activated sludge and chlorination process for disinfection.

Fecal coliforms are one of the indicators for health risk related bacteria in TRSE, and they were only detected a few times with the maximum value of 500 cfu/100 mL during the above experimental period. This indicates that the water quality of the observed TRSE belongs to Category B of ISO 16075-2, and it is suitable for irrigating crops that are eaten only after cooking. Additionally, TRSE has the potential to irrigate crops that require higher safety standards such as the crops that are eaten raw by contriving ways that prevent direct contact between irrigation water and crops, according to the ISO guidelines. Hence, TRSE is potentially suitable even for direct irrigation.

13.6 Summary

From the results of our test cultivation, the water quality of TRSE is suitable for agricultural reuse, and its negative effect on soil environments would be insignificant, if any. Further, it is even suitable for reuse in greenhouses. Positive reuse of TRSE, provided it is supplied close to farmlands, will secure a comparatively stable and safe water resource. Utilizing TRSE for irrigation purposes has the potential to be one of the solutions for water-stress improvement and aquatic environmental conservation around rural areas.

Densely populated area in Central Asia (i.e., Ferghana basin) is required for using TRSE and it is suitable for agricultural reuse. Nowadays, considering growing number of greenhouses in Central Asian countries, therefore it is even useful for reuse in greenhouses.

Acknowledgements This work was financially supported by "Pilot Program of International Collaborative Research (collaborative research based on a joint call with Israel)" under "Commissioned projects for promotion of strategic international collaborative research" funded by MAFF, and JSPS KAKENHI Grant Number JP17K08014.

References

1. Japan Heterological Agency. https://www.jma.go.jp/jma/ (in Japanese)
2. ISO (2015) ISO 16075(1-3)—Guidelines for treated wastewater reuse for irrigation

3. ISO (2016) ISO 16075(4)—Guidelines for treated wastewater reuse for irrigation
4. WHO (2006) WHO guidelines for the safe use of wastewater, excreta and greywater vol 2, Wastewater Use in Agriculture
5. Ministry of land, infrastructure of Japan, transport and tourism. https://www.mlit.go.jp/ (in Japanese)
6. Cowan PJ (2007) Geographic usage of the terms middle Asia and central Asia. J Arid Environ 2:359–363 [CrossRef]
7. Severskiy IV (2004) Water-related problems of central Asia: some results of the (GIWA) international water assessment program. Ambio 33:52–62 [CrossRef] [PubMed]
8. Tussupova K, Berndtsson R, Bramryd T, Beisenova R (2015/2016) Investigating willingness to pay to improve water supply services: application of contingent valuation method. Water 7:3024–3039. 8:219 10 of 13. [CrossRef]
9. Siegfried T, Bernauer T, Guiennet R, Sellars S, Robertson AW, Mankin J, Bauer-Gottwein P, Yakovlev A (2012) Will climate change exacerbate stress in central Asia? Clim Chang 112:881–899 [CrossRef]
10. Ministry of agriculture, forestry and fisheries. https://www.maff.go.jp/ (in Japanese)

Chapter 14
Designing Sustainable Futures: Interdisciplinary Science and Social Creativity

Ilan Chabay

Abstract In every region and every nation in the world, people face the difficult challenges of learning to design and maintain healthy societies with limited natural resources—including water, land, and energy—under rapidly changing global to local conditions. These challenges are contained within the framework of the 17 United Nations Sustainable Development Goals (SDGs) for 2030. Responding to the challenges of the SDGs at local to global scales must address the fundamental complexity of social–ecological systems. This requires understanding the technical, ecological, and geological conditions at multiple scales of time and space and equally importantly, understanding and designing for the future in the appropriate social, political, economic, and cultural contexts. In this talk, I want to discuss two critical questions that arise from these complex systems challenges: (1) How can we build the capacity to innovate for societal well-being by learning to think and act across traditional disciplinary lines? (2) How can we enhance the value of technology and engineering by engaging the knowledge, creativity, and cooperation of all parts of society in designing for their future?

Keywords Resource governance · Complex system science · Interdisciplinary research · Transdisciplinary research · Innovation for societal well-being · Inquiry and experiential learning

14.1 Introduction

The theme of this conference on "Current and Future State of Water Resource Management and Environmental Issues in Central Asia" is part of a larger set of challenges to the sustainability and quality of life for all of human society due to rapid and profound changes in the resources and bio-geo-physical, economic, and social conditions on Earth.

I. Chabay (✉)
Institute for Advanced Sustainability Studies, 130 Berliner Str., Potsdam 14467, Germany
e-mail: ilan.chabay@iass-potsdam.de

Arizona State University, School of Sustainability, Washington, DC, USA

Z. E. Kulenbekov and B. D. Asanov (eds.), *Water Resource Management in Central Asia and Afghanistan*, Springer Water, https://doi.org/10.1007/978-3-030-68337-5_14

The changes in the conditions for life on Earth comes not only from natural cycles and phenomena that we are familiar with throughout Earth's known history, but now are increasingly driven by human actions across the world at multiple temporal and spatial scales. We have to look at water resources, as well as energy, air pollution, climate change, food production, and land use changes, ocean temperature and acidification, and many others as critical human-influenced challenges to the necessary conditions for the well-being and even survival of human society. This profound shaping of the environment and conditions on Earth by humanity as active agents was represented as the "Anthropocene Era," a term coined by Paul Crutzen [1] and actively debated by stratigraphers and scientists ever since. Compelling evidence for this recent departure from the past patterns of human interaction with the environment can be seen in the many examples of the rapid onset and steep increase, particularly since the 1950s, in both earth system trends and socio-economic trends in the so-called "hockey stick" graphs [2].

The consequences of human impact and resource extraction in the Anthropocene Era have led to the recognition of the limitations of certain vital resources and living conditions on Earth. These were highlighted in the "Planetary Boundaries" papers [3, 4] in which the authors argue that for example, nitrogen and phosphorus flows, ocean acidification, land system changes, and biodiversity are at or approaching potential irreversible tipping points beyond which the consequences for human society could be devastating.

In response to the accumulating evidence of the high risk of undermining the conditions necessary for the long-term sustainability of human society, the United Nations adopted the 17 Sustainable Development Goals (SDGs) in 2015 [5]. While there are many specific targets under each of these goals, the SDGs are national and global aspirations, not mandatory directives.

The adoption of the goals was important as a step toward developing new strategies for moving toward sustainable futures in the diverse cultures and contexts of human societies. However, the 17 SDGs cannot be fulfilled in a linear managerial process, because they are inextricably interdependent in many cases and more generally, because the challenges of the Anthropocene Era are fundamentally due to the fact that human society is a complex system embedded in the complex bio-geo-physical system of Earth. Such complex systems are not amenable to traditional reductionist scientific approaches. They require new, collaborative interdisciplinary, and transdisciplinary methods and perspectives [6].

The topic of this conference, "Current and Future State of Water Resource Management and Environmental Issues in Central Asia", clearly identifies the context in which the relevant SDGs must be addressed in this part of the world. Goal 6—"Ensure availability and sustainable management of water and sanitation for all"—is obviously directly in line with the topic of the conference. Well before the UN agreement for the SDGs, the topic has been discussed for the post-Soviet era of independent republics in terms of water management [7]. But the water management issues in Central Asia are interdependent with other systems that are identified in other SDGs. For example, Goal 7—Ensure access to affordable, reliable, sustainable, and modern energy for all, Goal 13—Take urgent action to combat climate change and its impacts, ongoing research points to this interdependence and the difficult technical, social, and environmental challenges that Central Asia (and other parts of the world, as well) face in avoiding, mitigating, or adapting to global changes at multiple spatial and temporal scales [8, 9].

The future of water resource management and environmental in Central Asia is a very important example of the profound challenges of complex systemic risks to the ecosystems and social systems of the world. I want to discuss how academic scientists and engineers in collaboration with practitioners and civil society can approach the challenges in more effective ways.

14.2 Systems, Sustainability, and Society

In order to explore the strategies needed to improve the capacity of science to tackle these complex problems, the Global Sustainability Strategy Forum (GSSF), was founded in 2018 as a joint project of the Arizona State University in the US and the Institute for Advanced Sustainability Studies (IASS) in Potsdam, Germany, with funding from the Volkswagen Foundation. In the first of four GSSF events, 17 experts in sustainability science from different parts of the world met in March 2019. The five days of intense discussion focused on what changes in the social and natural sciences and in the education of scientists and society are needed to build the capacity to address the complex system challenge of sustainability more effectively.

The most important conclusions from the deliberations of the expert panel in the global sustainability strategy forum were that science and society must begin designing for long term desirable change, rather than focusing on maintaining or returning to prior patterns of living that seemed stable, but were unsustainable. That means there must be a greater focus on understanding social dynamics driving us towards unsustainable patterns of living, rather than being concerned primarily with mitigating the environmental consequences of unsustainable behaviors. This requires using complex systems science to address the sustainability challenges of socio-environmental systems [10]. It also requires learning from the past to understand the present and to imagine a more positive future. It is important here to emphasize the necessity of using imagination and the ability to go beyond linear extrapolation and path dependence to be able to emerge successfully from our current conundrum [11].

These changes in science require a change in our educational strategies to build the capacity for transformative interdisciplinary and transdisciplinary research for sustainability. We need the capacity to do complex systems research for the needs of present and future societies. In addition to the understanding of complex system science, we need to develop the process of co-designing research and co-producing knowledge with diverse knowledge holders and stakeholders in the society [12–14]. Scientists need to engage with the creativity and insights of their fellow citizens of Earth in open mutual learning dialogues with all sectors of society for developing new practices and policies. Dealing effectively with the issues of water management and environment in Central Asia requires the depth of local knowledge of the hydrological, social, economic, and environmental details in Central Asia. Excellent generic knowledge and global perspectives are insufficient to find creative, robust, and meaningful solutions in the cultural and contextual differences of different regions. This is the basis for engaging with local knowledge holders and stakeholders in transdisciplinary research [15, 16].

Fundamental to science in its relationship with society is the recognition that society is deeply embedded within the natural systems on which it is entirely dependent. The challenges of avoiding transgression of the planetary boundaries and overdrawing limited resources are rooted in societal challenges, rather than solely environmental or technological ones. This follows from the observation that societies each define what is relevant and valuable in their relationship to the local and global environment [17]. Consequently, the central issue is not simply the bio-physical or technological conditions in a particular place, but how the people in that location or community view their environment and circumstances.

If members of a community think that some intervention in the conditions of their community are needed, who has the authority, agency, and responsibility to act? Do individuals feel that they have agency, knowledge, and responsibility to act? Agency means that an individual (or the members of a group) is capable of doing something and has the necessary resources with which to act. If the person or group is capable of acting, do they feel responsible for acting? In other words: can I act and should I act?

However, acting for a sustainable future is not a single intervention at one point in time. It will require a process of learning how to act individually and collectively as conditions and challenges change. We must continually learn and co-design and innovate to meet the changing needs of societies in moving toward sustainable futures in which the well-being of all is taken into account in a just and equitable way.

Thus, a key challenge is changing to sustainable collective human behaviors, rather than focusing on mitigating the consequences of unsustainable behaviors. Addressing the challenge of the collective behavior change is the mission of the Knowledge, Learning, and Societal Change Alliance (KLASICA). KLASICA examines how knowledge, meaning-making, and collective behavior change are interconnected [18] by identifying and understanding the conditions under which collective behavior change toward sustainable futures occurs in the contexts of different communities and cultures. We are learning from narrative expressions of visions of the future and of social identities to understand better the dynamics of societal movements toward sustainable futures with regard to specific needs in different communities around the world [19]. The understanding, including insights gained by building models of social dynamics is being developed to find new solutions for effective actions on the pathways to sustainable futures.

14.3 Knowledge, Creativity, and Innovation for Society

Science and technology contribute knowledge to society that allows for changes in patterns of behavior and policy, but the challenge is to make those changes constructive in leading toward sustainable futures, rather than to fall victim to unintended consequences and technological fixes which may only exacerbate the problems. We need more than good technology and science by themselves. We need imagination and creativity and innovation for the needs of society. We must learn from the past to understand our current situation and to imagine a desirable future that provides for the well-being of our societies, including the futures of our children and their children and beyond. To do this we need to focus on the systemic risks of deeply entwined nature and society. We need to break from the path dependencies in our thinking and find other ways to look ahead. Can we imagine the future and think backward to find a new path forward, rather than thinking from the present toward the future in familiar incremental steps. To do this, scientists must engage collaboratively with diverse thinkers and knowledge holders to expand the creativity needed to innovate for the well-being of society. We need less linear and more associative pattern-sensitive thinking. We need new models to help expand our thinking, not only to try to predict future developments from those existing patterns. And we need to develop a continually adaptive governance process that makes use of science as a resource for evidence and processes responsive to the needs of policymakers [20, 21].

Fostering the engagement between scientists and society in new ways will require new approaches to learning across all of the ages of humanity from early childhood to lifelong learning. That's essential because we need to change the expectations in society about learning and about what science offers in helping to shape our future. This process of learning at all ages really should begin by stimulating curiosity and questions from learners at any age, rather than starting by handing out information known by experts. In many places around the world, this is not familiar as a way of learning. Teaching is practiced as a top-down process of imparting specific knowledge in traditional forms. Yet all children exhibit the curiosity that can open this process up for them and allow them to engage throughout their lives with both curiosity and mastery of essential ideas and skills (e.g., see [22, 23]). Teachers are familiar and comfortable with the way they were taught. So reforming teacher education in different cultures and circumstances is a challenging process [24]. Introducing learner-centered and project-based learning can contribute substantially to education that better enables societies to creatively innovate and adapt to changing circumstances around us.

The motivation to learn and to look more openly at the challenges of the world around us can only be done by stimulating and rewarding curiosity and by building relevant experience. Providing experiences with phenomena and asking deeper questions is part of the building of vocabulary for thinking that is based in knowledge of how things work. As part of this, we need to make the use of models an essential part of the learning process [25, 26]. There is a common practice of using results of models as the answers, rather than learning from the actual construction and use of models to create understanding. Models provide a way to safely explore new ideas and test assumptions. They are also a means for engaging people in exploring ideas as scenarios for discussion and thereby facilitating dialogue on complex issues including societal issues. Physical and electronic models and games can serve as important boundary objects that facilitate dialogue among diverse participants [27–29].

14.4 Engaging Society in Games, Exhibitions, and Dialogues

Models, games, and public exhibitions are important for engaging with the broader segment of society to support their understanding of key ideas for sustainability. Learning to engage in transdisciplinary dialogue with diverse members of society is essential to address complex issues that are rooted in the value systems of the society. That allows scientists to reach more broadly into the knowledge base and value systems for more effective use of the knowledge both of science and from society. Figures 14.1 and 14.2 illustrate transdisciplinary dialogues. In Fig. 14.1, the dialogue involved stakeholders and rights-holders from civil society, business, and governments on Arctic sustainability concerns.

The two examples of dialogues were undertaken for different reasons, but they have in common that they engaged participants in a discussion to begin to understand their different perspectives and decide how they can use their own knowledge and science to achieve better outcomes for their communities.

Figures 14.3 and 14.4 are examples of highly engaging exhibitions and toys designed by the author and used to stimulate open dialogues with people about their questions, ideas, and concerns.

Fig. 14.1 The image is of stakeholder and rights-holders in a dialogue organized by the author and colleagues at the Arctic Circle meeting in Reykjavik, Iceland in 2014 (photo by Ilan Chabay)

Fig. 14.2 A community meeting on earthquake recovery guided by a local Buddhist monk following a major earthquake near Nishihara, Kyushu Japan in 2016 (photo by Ilan Chabay)

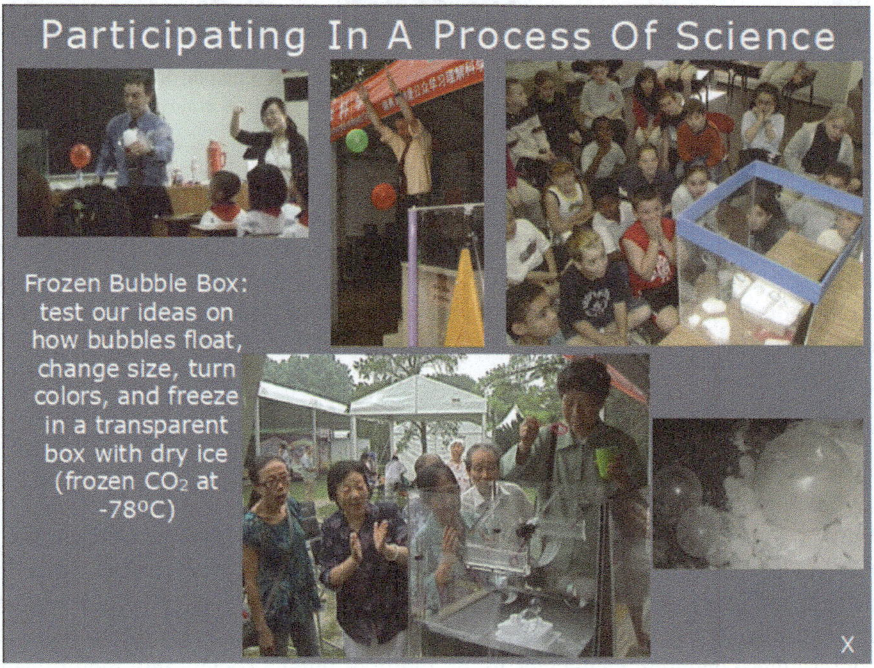

Fig. 14.3 A participatory activity called "Frozen Bubble Box" that simulates great curiosity and opens discussions on the process of doing science, on global warming, ocean acidification, and many other topics of chemistry, physics, and biology (photo by Ilan Chabay)

Fig. 14.4 Four images of an exhibit and game designed by the author that involves the production of oxygen and hydrogen from water as a clean fuel and illustrates in a visceral way the relationship between work, power, and energy (photo by Ilan Chabay)

Figure 14.5 shows a mobile exhibition used in Germany to engage people with questions about renewable energy and sustainability. In the center of the exhibits was a game that was played for only a short time (3–5 min), but which was highly memorable and provided a starting point for discussions on energy and renewable resources [30].

14.5 Conclusion

While the theme of this conference is water resource management in Central Asia, the wider context that I outlined here is that of the challenge of sustainability in different contexts and cultures. Managing or better yet governing resources are part of the challenge that the changes on Earth Pose for humanity. Finding new ways to address these daunting challenges to the future of human society will require new approaches to science, to learning, and to engaging between science and society. We will have to find the will and the way to develop collective behavior change based on scientific knowledge in combination with local knowledge and culturally traditional

Fig. 14.5 A mobile exhibition in Germany on renewable energy and sustainability (photo by Ilan Chabay)

knowledge. This will help different communities to find their own desirable future within the global coherence needed for the well-being of all societies. That is the fundamental challenge we must face and move forward in doing this together.

References

1. Crutzen PJ (2002) Geology of mankind. Nature [Internet]. [cited 2019 Mar 19]; 415:23–23. Available from: https://www.nature.com/articles/415023a
2. Steffen W, Broadgate W, Deutsch L, Gaffney O, Ludwig C (2015) The trajectory of the anthropocene: the great acceleration. Anthr Rev 2:81–98
3. Rockström J, Steffen W, Noone K, Lambin E, Lenton TM, Scheffer M et al (2009) Planetary boundaries: exploring the safe operating space for humanity. Ecol Soc. 14:32
4. Steffen W, Richardson K, Rockstrom J, Cornell SE, Fetzer I, Bennett EM et al (2015) Planetary boundaries: guiding human development on a changing planet. Science [Internet]. [cited 2019 Mar 19]; 347:1259855–1259855. Available from: https://doi.org/10.1126/science.1259855
5. United Nations General Assembly (2015) Transforming our world: the 2030 agenda for sustainable development. https://sustainabledevelopment.un.org/content/documents/7891Transforming%20Our%20World.pdf. pp 1–5
6. van der Leeuw SE (2014) Sustainability, culture and personal responsibility. Sustain Sci [Internet] 9:115–117. https://doi.org/10.1007/s11625-014-0249-5

7. O'Hara SL (2000) Lessons from the past: water management in Central Asia. Water policy [Internet]. 2:365–84. Available from: https://linkinghub.elsevier.com/retrieve/pii/S1366701700000106
8. Deng H, Chen Y (2017) Influences of recent climate change and human activities on water storage variations in Central Asia. J Hydrol [Internet]. Elsevier B.V.; 544:46–57. Available from: https://doi.org/10.1016/j.jhydrol.2016.11.006
9. Reyer CPO, Otto IM, Adams S, Albrecht T, Baarsch F, Cartsburg M et al (2017) Climate change impacts in Central Asia and their implications for development. Reg Environ Chang [Internet]. 17:1639–50. Available from: https://doi.org/10.1007/s10113-015-0893-z
10. Bogardi J, Chabay I, Kerekes S (2019) Uncertainty and complexity: scientific challenges in understanding the unsustainable present and preparing for an unknown future. In: Jensen J (ed) Risk, Responsibility, and Resilience—new strategies in the labyrinths of uncertainty. Institute for Advanced Studies, Közseg Hungary, pp 87–101
11. Bai X, Begashaw B, Bursztyn M, Chabay I, Droy S, Folke C et al (2019) Changing the scientific approach to fast transitions to a sustainable world. Improving knowledge production for sustainable policy and practice [Internet]. Available from: https://publications.iass-potsdam.de/pubman/item/escidoc:4408890:3/component/esc
12. Reed MS, Stringer LC, Fazey I, Evely AC, Kruijsen JHJ (2014) Five principles for the practice of knowledge exchange in environmental management. J Environ Manage. 146
13. Pahl-Wostl C, Tàbara D, Bouwen R, Craps M, Dewulf A, Mostert E et al (2008) The importance of social learning and culture for sustainable water management. Ecol Econ 64:484–495
14. Foley RW, Wiek A, Kay B, Rushforth R (2017) Ideal and reality of multi-stakeholder collaboration on sustainability problems: a case study on a large-scale industrial contamination in Phoenix, Arizona. Sustain Sci 12
15. Ely A, Marin A, Charli-Joseph L, Abrol D, Apgar M, Atela J et al (2020) Structured collaboration across a transformative knowledge network—learning across disciplines, cultures and contexts? Sustainability 12:2499
16. Mielke J, Vermaßen H, Ellenbeck S (2017) Ideals, practices, and future prospects of stakeholder involvement in sustainability science. Proc Natl Acad Sci [Internet]. 201706085. Available from: https://doi.org/10.1073/pnas.1706085114
17. TWI2050—The World in 2050 (2018) Transformations to achieve the sustainable development goals—report prepared by the world in 2050 initiative [Internet]. Int Inst Appl Syst Anal Available from: https://pure.iiasa.ac.at/15347
18. Chabay I (2019) Narratives, networks, and knowledge: finding pathways to sustainable futures. In: Jensen J (ed). Risk, responsib resil new strateg labyrinths uncertain. Közseg, Hungary: insitute for advanced studies, Köszeg Hungary. pp 77–86
19. Chabay, I, Koch, L, Martinez, G, Scholz, G (2019) Influence of narratives of vision and identity on collective behavior change. Sustainability [Internet]. [cited 2019 Oct 27]. 11:5680. Available from: https://www.mdpi.com/2071-1050/11/20/5680/pdf
20. Clark WC, van Kerkhoff L, Lebel L, Gallopin GC (2016) Crafting usable knowledge for sustainable development. Proc Natl Acad Sci [Internet]. [cited 2017 Aug 8]. 113:4570–8. Available from: https://doi.org/10.1073/pnas.1601266113
21. Pahl-Wostl C (2009) A conceptual framework for analysing adaptive capacity and multi-level learning processes in resource governance regimes. Glob Environ Chang 19:354–365
22. Borg F, Winberg M, Vinterek M (2017) Children's learning for a sustainable society: influences from home and preschool. Educ Inq Routledge 8:151–172
23. Ortega-Sánchez D, Jiménez-Eguizábal A (2019) Project-based learning through information and communications technology and the curricular inclusion of social problems relevant to the initial training of infant school teachers. Sustainability [Internet]. [cited 2019 Nov 23]. 11:6370. Available from: https://www.mdpi.com/2071-1050/11/22/6370
24. Carter L (2008) Sociocultural influences on science education: innovation for contemporary times. Sci Educ pp 165–81
25. Amsler S, Facer K (2017) Contesting anticipatory regimes in education: exploring alternative educational orientations to the future. Futures 94:6–14

26. Voinov A, Kolagani N, McCall MK, Glynn PD, Kragt ME, Ostermann FO et al (2016) Modelling with stakeholders—next generation. Environ Model Softw [Internet]. Elsevier. [cited 2018 Aug 23]. 77:196–220. Available from: https://www.sciencedirect.com/science/article/pii/S1364815215301055
27. Chabay I (2018) Taking time, sharing spaces: adaptive risk governance processes in rural Japan. Int J Disaster Risk Sci [Internet]. Beijing Normal University Press. [cited 2018 Nov 10]. 9:464–71. Available from: https://doi.org/10.1007/s13753-018-0191-8
28. Roux DJ, Nel JL, Cundill G, O'Farrell P, Fabricius C (2017) Transdisciplinary research for systemic change: who to learn with, what to learn about and how to learn. Sustain Sci 12:711–726
29. Louca LT, Zacharia ZC (2012) Modeling-based learning in science education: cognitive, metacognitive, social, material and epistemological contributions. Educ Rev 64:471–492
30. Li H, Chabay I, Renn O, Weber A, Mbungu G (2015) Exploring smart grids with simulations in a mobile science exhibition. Energy Sustain Soc [Internet]. 5:1–8. Available from: https://doi.org/10.1186/s13705-015-0066-4

Part V
Application of GIS and Remote Sensing in Water Resources Management

Chapter 15
Investigation of the High Mountain Vegetation Using Satellite Imagery, Kyrgyzstan

Zheenbek E. Kulenbekov, Sagynbek Zh. Orunbaev, and Baktyiar D. Asanov

Abstract This study uses GIS and remote sensing applications to investigate land cover changes in the Alay Valley, Osh oblast, Kyrgyzstan over time using advanced methods such as NDVI calculation. It is concluded that NDVI fluctuates seasonally. According to the Pearson correlation coefficient analysis between NDVI and precipitation over the given period of time, very surprisingly, correlation turned out to be very weak, only 0.283, with a P-value equal to 0.241. However, Pearson correlation coefficient analysis between NDVI and temperature over the same time period was very high, 0.915, with a P-value equal to 0.000 indicating very strong correlation. Visually from maps, it is possible to judge how productive and healthy croplands are. The highest values and the lowest point of NDVI were found for over years, i.e., 1993–2017. From this data it is also possible to draw the conclusion that the highest month in terms of NDVI is August and lowest is May.

Keywords NDVI · Correlation · Precipitation · Temperature · Interrelationship

15.1 Introduction

Land cover information plays significant role in climate change studies as well as in understanding complex interconnectivities between human activities and global change. Thus, accurate and up-to-date land cover information plays critical role in sustainable resource management, planning and monitoring activities [1]. Conventional methods of land cover classification, generally constrained by field surveys, turn out to be insufficient and time-consuming. Yet, despite the fact remote sensing has long served as a generator of accurate land cover information, problems still occur when working with extraction of land cover information due to several reasons [2]. Improper and mishandled satellite image treatment may result in inappropriate output in classification of land cover. Much work has been recently done on climate change vegetation response to identify spatio-temporal variations in Central Asia. Earth soil

Z. E. Kulenbekov (✉) · S. Zh. Orunbaev · B. D. Asanov
American University of Central Asia, 7/6 Aaly Tokombaev Str., Bishkek, Kyrgyzstan

Z. E. Kulenbekov and B. D. Asanov (eds.), *Water Resource Management in Central Asia and Afghanistan*, Springer Water, https://doi.org/10.1007/978-3-030-68337-5_15

has always been subject to change and alteration due to both natural and anthropogenic factors. However, over the last couple of decades, Central Asia has suffered from severe land cover and land use alterations caused partly by socio-economic and institutional changes after the collapse of the Soviet Union. Central Asia encompasses the largest adjoined area of rangeland worldwide [3], but has low capacity to properly manage it [4].

As noted above, the complexity inherent in changing land resource cover and use includes both natural and human factors; these include factors such as unsustainable pasture management that leads to land degradation due to overgrazing and natural phenomena such as salinization of soils, desertification, and erosion. In the mountainous ecosystems of Kyrgyzstan, which are highly vulnerable to natural and anthropogenic impacts, biological resources are being depleted at an increasing rate, and soil erosion and decrease in land productivity have already reached a critical point [5].

Based on climate change scenario estimation, Kyrgyzstan will face drastic annual and seasonal variations of temperature and precipitation [6]. Some valuable ecosystems will thus be threatened and are subject to loss due to both unsustainable natural resources use and climate change [4, 7]. Therefore, a study on how the influence of temperature and precipitation can change vegetation dynamics is necessary for better land use economy and climate change adaptation planning.

Livestock husbandry and agriculture are essential sectors of the Alay and Chon-Alai Region in Southern Kyrgyzstan and play critical roles in the local economy [4]. Animal husbandry is highly dependent on climate conditions which have a great impact on pastureland and fodder. Periodic natural disasters such as landslides and earthquakes, as well as heavy precipitation and long winters, severely damage local economies.

15.1.1 Research Questions and Objectives

This study aims to identify seasonal variations in the normalized difference vegetation index (NDVI), precipitation, and temperature in the Alay Valley. In addition, the research focuses on finding an interrelationship between the abovementioned variables. The following questions are answered in this research paper:

Is there any seasonal interrelationship between NDVI, temperature, and precipitation in the Alay Valley?

If there is some interrelationship between these parameters, is it possible to predict one of them knowing the other?

How did NDVI values change over time from 1993 to 2003?

Subsidiary question:

Is there any significant NDVI change over that last 5 years? (Fig. 15.1).

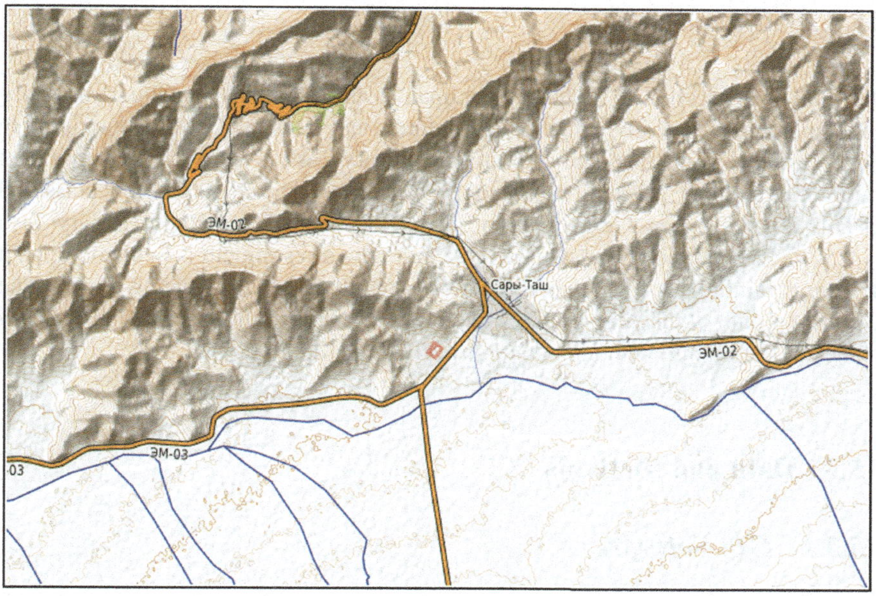

Fig. 15.1 Topographic map of Sary-Tash [https://kac.centralasia.kg/maps/#topographical-maps]

15.2 Study Area

15.2.1 Study Area

The study area covers the Alay Valley district located on the southwestern part of Kyrgyzstan. The area of particular interest is the Sary-Tash Village; coordinates: 39°43′48″N 73°15′00″E with total area of 6.41 km^2.[1] Sary-Tash is conveniently situated at the convergence of three roads and makes a good place to break up the Murgab (Tajikistan) to Osh (Kyrgyzstan) or Kashgar (China) to Osh trip as it sits at the crossroads of the Pamir Highway and the road from nearby China to Tajikistan through the Alay Valley. The study area embraces various ecological zones and biomes including steppes, hills, mountains, rocks, valleys, and grasslands as well as man-managed agricultural lands and pasture rangelands, at an elevation of around 2500–3000 m.a.s.l [8].

[1] https://www.maplandia.com/kyrgyzstan/osh/sary-tash-39-44-0-n-73-15-0-e/villages/.

15.2.2 Climate

The Alay Valley climate is sharply continental and is subject to annual variations in temperature—severely cold and long winters and moderately warm summers. In Sary-Tash, the climate is cold and temperate. Sary-Tash is a city with a significant rainfall. Even in the driest month, there is a lot of rain. The climate here is classified as Dfb[2] by the Köppen–Geiger system. The average annual temperature in Sary-Tash is 0.5 °C. Precipitation here averages 322 mm. The driest month is September, with 12 mm of rainfall. Most of the precipitation here falls in May, averaging 49 mm. The warmest month of the year is July, with an average temperature of 14.0 °C. January is the coldest month, with temperatures averaging −15.5 °C.

15.3 Data and Methods

15.3.1 GIS Software

GIS software embraces a wide range of applications, including data visualization, geocoding, and spatial querying. GIS software is either open source and free or commercial, and can be classified into several categories of GIS: Desktop, Quantum, SAGA, GRASS, TerraView, FalconView, ILWIS, and so on.

There are a variety of software applications and tools used for LCLU (land cover and land use classification). Users can choose any program for classifying specific areas depending on project goals and available resources. A plenty of open source, free-to-download software is being used globally for LCLU. However, sometimes commonly used free source software is limited functionally, for example, some commercial GIS software such as ArcGIS, ESRI, and ENVI have numerous high-level geospatial and geostatistical tools that are missing in open-source software, Quantum, GRASS, and SAGA GIS.

15.3.2 Correlation

As a rule, correlation coefficients are used in statistics to measure interdependence between two variables. These can be either quantitative variables or categorical variables[3] and generate a probabilistic or statistical dependence, which, generally speaking, does not have a strictly functional character. Unlike functional dependence, correlation dependence arises when one of the characteristics depends not

[2]https://en.wikipedia.org/wiki/Humid_continental_climate.

[3]Stephanie. "What Is Correlation in Statistics? Correlation Analysis Explained." Statistics How To, 10 Aug. 2013, www.statisticshowto.com/what-is-correlation/.

only on the given second, but also on a number of random factors, or when among the conditions on which both signs depend, there are common conditions for both of them. When calculating correlations, one tries to determine whether there is a statistically reliable relationship between two or more variables in one or more samples. It is important to understand that correlation dependence reflects only the relationship between the variables and does not speak of cause–effect relationships. The correlation relationship only indicates the interconnectedness of these parameters in the particular sample under study; in another sample, we may not observe the initial correlations obtained. The correlation coefficient (r) characterizes the value reflecting the degree of interrelation of two variables among themselves. It can vary from -1 (negative correlation) to $+1$ (positive correlation). If the correlation coefficient is 0 then this indicates that there are no correlations between the variables. Thus, if the correlation coefficient is closer to 1 (or -1), then there is a strong correlation, and if closer to 0, then it is weak. With positive correlation, an increase (or decrease) in the values of one variable leads to a regular increase (or decrease) in the other variable, i.e., interrelationships such as increase–increase (decrease–decrease). The inverse is true with regard to negative correlation.

15.3.3 Landsat's Land Cover Classification Method

Land cover classification and land surface monitoring through the Landsat satellite program have a rich history. Landsat 1 was launched in July 1972, and since then several methods for land classification have been incorporated. Land cover classification using Landsat satellite images has become an important matter of discussion in the context of climate change over the last couple of decades [9]. Landsat satellite imagery has continuously been advancing with the development of newly upgraded sensors. This is mainly achieved by the upgrading the level of spatial, spectral, radiometric, and temporal resolution [10].

It is worth mentioning that, relatively recently, Landsat images have become integrable with other satellite images through the image fusion technique. The invention of new remote sensing technologies such as advanced very-high-resolution radiometer (AVHRR) and moderate-resolution imaging spectroradiometer (MODIS) as well as advancement in image fusion methods have dramatically increased the quality of land cover classification [9].

15.3.4 Methods

Despite the fact that the study area is located in southern part of Kyrgyzstan, the vegetation period starts far later than in the north of the country due to the geographical and climatic conditions, i.e., the altitude and long, cold winters. For that reason, NDVI values were calculated from May to September to trace the dynamics of vegetation growth. Landsat 5 TM was used for years 1993–1998. Landsat 7 ETM+ satellite imagery was employed for calculation of NDVI from 1999 to 2003. Sentinel 2 satellite images (Satellite Platform: S2A_*; Product Type: S2MSI1C) were used for the period from August 2017 to June 2018. For the raster image processing, i.e., NDVI calculation, QGIS 3.4.2 software was used. The same application was used for land cover classification using the Semi-Automatic Classification Plugin. As for data visualization, Minitab-18 statistical software was used to plot scatterplots, pie charts, graphs, and other visual representations of the results. Satellite images from Landsat 7 ETM+ and Landsat 8 OLI were atmospherically and radiometrically corrected.

15.3.5 NDVI Calculations

Landsat satellite images were downloaded from the United States Geological Survey website, particularly Landsat 5 Thematic Mapper (TM) and Landsat 7 Enhanced Thematic Mapper + (ETM) and Landsat 8 OLI. Sentinel 2 products were downloaded from European Space Agency Copernicus Open Access Hub website.[4] Both Landsat and Sentinel 2 products were imported into QGIS Desktop 2.18.18 and after some image preprocessing procedures, such as radiometric calibration, atmospheric correction, and conversion from the raw digital numbers (DN) to surface reflectance, NDVI values were calculated. The time periods covered by these data are from 1993 to 1996 and from 2000 to 2003. NDVI values for the period from August 2017 to June 2018 were calculated using Landsat 8 OLI (Operation Land Manager) and Sentinel 2 imagery. These satellite images were downloaded from the United States Geological Survey website before being processed in QGIS 2.18.18 with radiometric calibration, atmospheric correction, and conversion from digital number to surface reflectance. For different Landsat satellite products, different band combinations were used: for Landsat 5 TM as well as for Landsat 7 ETM+ , Band 3 (Red) and Band 4 (Near Infrared) were used, whereas for Landsat 8 OLI, Band 4 (Red) and Band 5 (Near Infrared) were imported. To calculate NDVI from Sentinel 2 product, Band 8 and Band 4 were used. Calculations were made in the Raster Calculator using previously mentioned NDVI formula, and maximum, mean, and minimum values were found.

15.3.6 Land Cover Classification

Figure 15.2 illustrates one of the algorithms for land cover classification. The procedure of land cover classification was similar to NDVI determination at the early stages, involving image preprocessing steps including radiometric calibration, atmospheric correction, and conversion from digital number to surface reflectance. The Semi-Automatic Classification plugin—an open-source plugin for QGIS for both

[4]https://scihub.copernicus.eu/dhus/#/home.

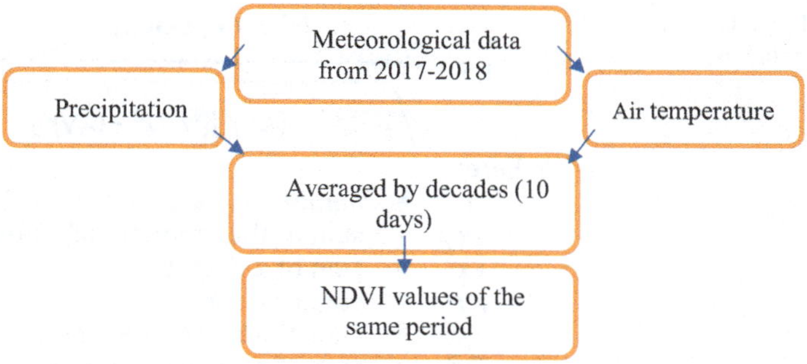

Fig. 15.2 Algorithm for finding the correlation coefficient

supervised and unsupervised classification analysis—was then used. To make the classification, all the 11 bands were necessary for Landsat 8 OLI. The supervised classification method was selected for as the authors had been to the site and had prior knowledge and a general idea of different "classes" in that region. For classification, it is necessary to create a Training Input File for identifying ROIs (regions of interest). This involves adding "Classes and Macroclasses," for example, a class can be river, whereas macroclass is water resources as a whole. ROIs were identified by highlighting polygons over the area of interest. Thus, to create a class of cropland, a polygon was drawn over agricultural fields. To make sure classification was of good quality, the classification preview regime was activated. Following the preview, the final step was to select an algorithm. The "Maximum Likelihood" algorithm was selected as it fits best the supervised method of classification in QGIS. Four different classes were generated for Sary-Tash: Cropland, Water, Bare Soil, and Vegetation.

15.3.7 Temporal Characteristics

Due to the long winters and short vegetation period, it was logical to calculate NDVI only during the period of vegetation—from May to September. In Sary-Tash, as in the whole Alay Region, the snow melts only by the end of April and at the beginning of May. This was the main reason behind the decision to start finding NDVI values from May. At the beginning of September, practically all the vegetation stops growing—this is why NDVI calculation ends at the end of September.

15.3.8 Meteorological Data

Meteorological data was obtained from the local weather station daily weather data, and it includes air temperature, land surface temperature, and precipitation, as well as other parameters that do not relate to this study.

Fig. 15.3 Formula for the pearson correlation coefficient

$$r = \frac{N\Sigma xy - (\Sigma x)(\Sigma y)}{\sqrt{[N\Sigma x^2 - (\Sigma x)^2][N\Sigma y^2 - (\Sigma y)^2]}}$$

Where:

N = number of pairs of scores
Σxy = sum of the products of paired scores
Σx = sum of x scores
Σy = sum of y scores
Σx^2 = sum of squared x scores
Σy^2 = sum of squared y scores

15.3.9 Statistical Analysis

One of the main goals of the study is to identify whether there is a correlation between NDVI and climatic variables. For that, the Minitab-18 statistical software application was used. All data from weather station was imported into the program and the average for every 10 days from May to November was calculated. NDVI values calculated in QGIS were then imported into Minitab as well. Once mean meteorological data values were found, the Pearson correlation coefficient between these and NDVI was calculated using the formula in Fig. 15.3.

15.4 Results and Discussion

15.4.1 Temporal Distribution of NDVI and Climatic Parameters

In Fig. 15.4, it can be seen that NDVI fluctuates seasonally; it reaches its peak in August and then smoothly decreases to the lowest point in November. The same pattern is evident with air temperature (Fig. 15.5)—the highest and lowest values are found in August and November, respectively. Precipitation amount is distributed in slightly different way (Fig. 15.6)—November and June are found to be the highest months in terms of precipitation.

15.4.2 Pearson Correlation Coefficient Analysis

In Fig. 15.7, the Pearson correlation coefficient analysis between NDVI and precipitation over the given period of time can be seen. Quite surprisingly, correlation turned out to be very weak—only 0.283 with a P-value equal to 0.241. However, Pearson correlation coefficient analysis between NDVI and temperature over the same time

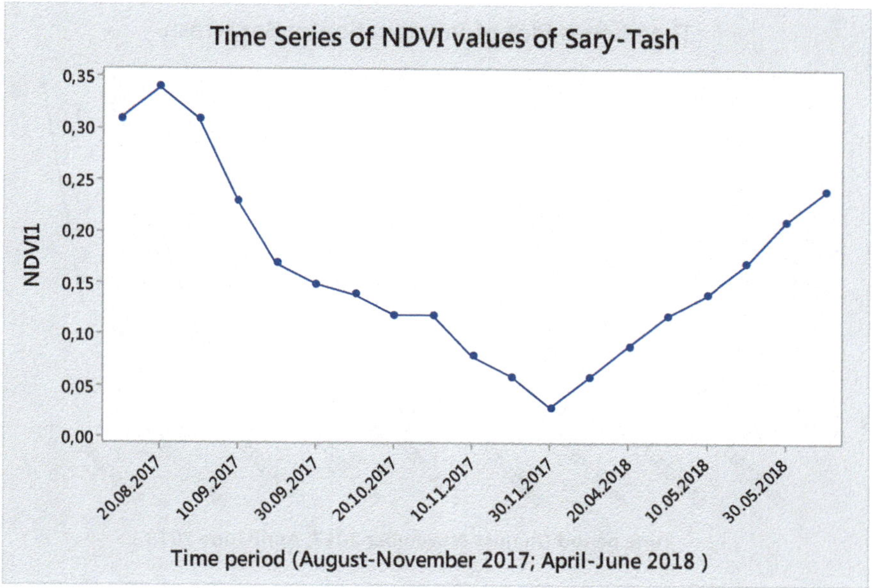

Fig. 15.4 Time-series plot of NDVI

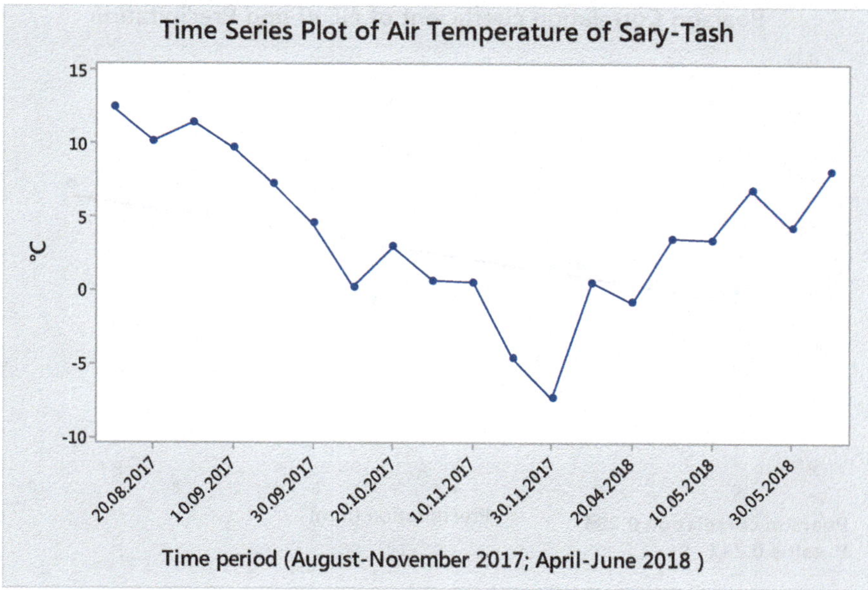

Fig. 15.5 Time-series plot of air temperature

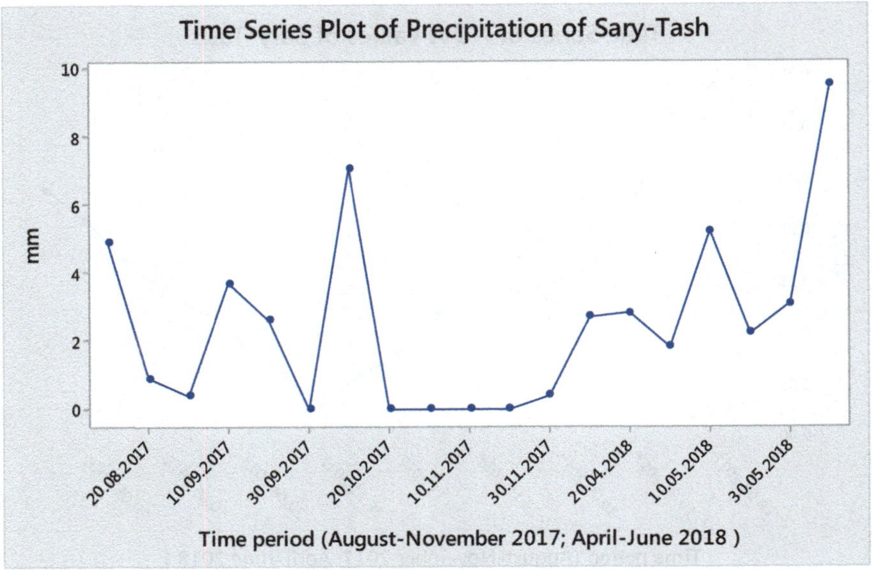

Fig. 15.6 Time-series plot of precipitation

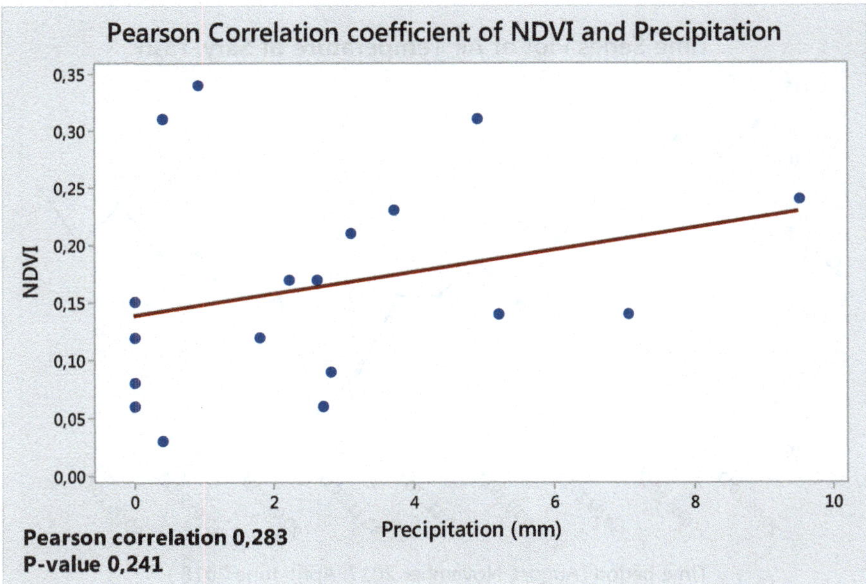

Fig. 15.7 Pearson correlation coefficient between NDVI and precipitation

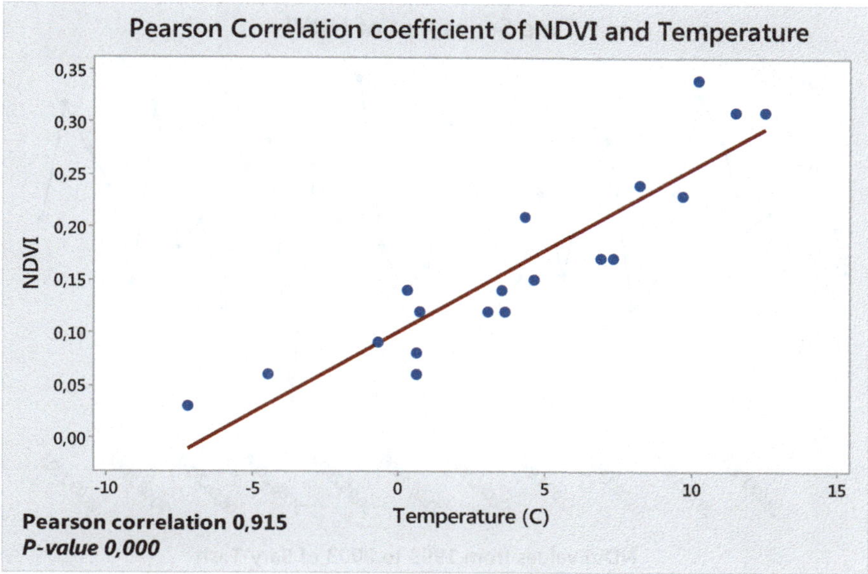

Fig. 15.8 Pearson correlation coefficient between NDVI and temperature

period (Fig. 15.8) was very high, 0.915, with a P-value equal to 0.000 indicating very strong correlation. Figure 15.9 represents a time-series plot of NDVI in Sary-Tash Village over the period of 1993 to 2003. The highest values were found in years 1993, 1999, and 2003 (0.38, 0.40, and 0.37, respectively). The lowest point was found in the year 2000—0 NDVI. From the data, it is also possible to draw the conclusion that the highest month in terms of NDVI is August and lowest is May.

Table 15.1a depicts the meteorological data by decade (August 2017–June 2018), while Table 15.1b represents NDVI values over the same period of time and Table 15.2 shows NDVI values of Sary-Tash for the 1993–2003 time period. Figures 15.10, 15.11, 15.12, 15.13 and 15.14 represent four different images—band combinations of Sary-Tash—which were made at the same period of time per year and can be interpreted as land cover classifications. The NDVI map allows for the generation of images that show active biomass and helps to make drought and agriculture production forecasts. For the NDVI map, bands 4 and 5 have been used; darker green color indicates high NDVI and white indicates low NDVI. Healthy vegetation maps illustrate band combinations of Landsat 7 ETM+ (Band 4—near infrared, Band 5—shortwave infrared, and Band 1—blue) and Landsat 8 OLI (Band 5—near infrared, Band 6—shortwave infrared, and Band 2—blue). Healthy vegetation can be depicted by band combination of Landsat 7 ETM+. Thus, it shows dense and healthy plants in green and light green; water bodies are dark (either violet or blue). Bare soil and rocks are represented in white, yellow, orange, red, and brown colors. Visually, it is possible to judge how productive and healthy croplands are. Overall, in agriculture maps, healthy vegetation looks bright green and is composed of bands 6, 5, and 2.

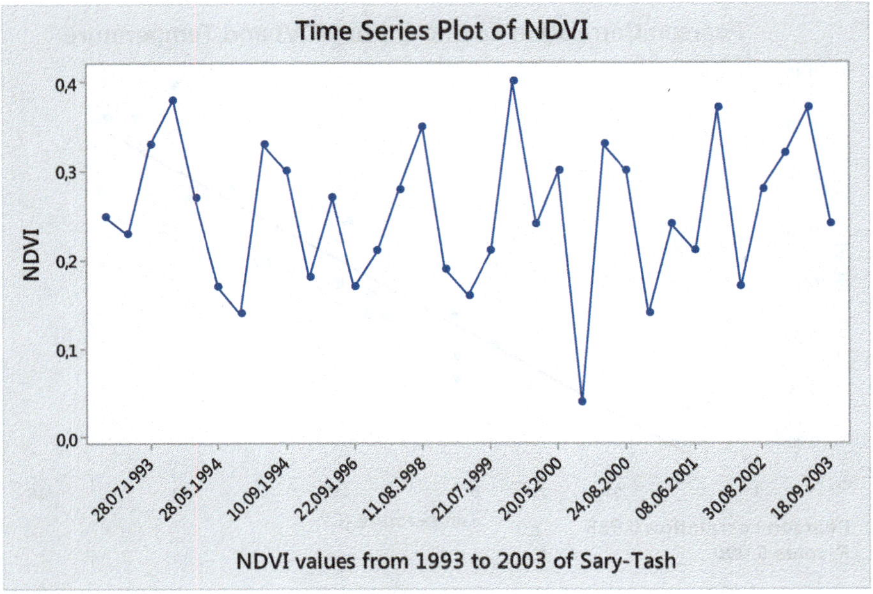

Fig. 15.9 Time-series plot of NDVI

Image stacks can be interpreted as land cover classification as well. In these, in fact, the vegetation is green; water is violet; clouds are purple; and soils, mountains, and bare rocks are blue. NIR, red, green, and SWIR 1 bands are used for this classification.

Figure 15.15 demonstrates time-series analysis of NDVI values over the last 5 years, and it is clear that there is no abnormality—no temporal variations except for the year 2017.

15.5 Conclusion

It is concluded that NDVI fluctuates seasonally. According to the Pearson correlation coefficient analysis between NDVI and precipitation over the given period of time, very surprisingly, correlation turned out to be very weak, only 0.283, with a P-value equal to 0.241. However, Pearson correlation coefficient analysis between NDVI and temperature over the same time period was very high, 0.915, with a P-value equal to 0.000 indicating very strong correlation. The highest values were found in years 1993, 1999, and 2003 (0.38, 0.40, and 0.37, respectively). The lowest point was found in the year 2000—0 NDVI. From the data, it is also possible to draw the conclusion that the highest month in terms of NDVI is August and lowest is May.

Table 15.1 **A** Average temperature and precipitation values and **B** NDVI values from August 2017 to June 2018

A			B	
Date	Temp	Precip	Date	NDVI
10.08.2017	12.5	4.9	10.08.2017	0.31
20.08.2017	10.2	0.9	20.08.2017	0.34
30.08.2017	11.5	0.4	30.08.2017	0.31
10.09.2017	9.7	3.7	10.09.2017	0.23
20.09.2017	7.3	2.6	20.09.2017	0.17
30.09.2017	4.6	0	30.09.2017	0.15
10.10.2017	0.3	7.05	10.10.2017	0.14
20.10.2017	3.02	0	20.10.2017	0.12
30.10.2017	0.7	0	30.10.2017	0.12
10.11.2017	0.6	0	10.11.2017	0.08
20.11.2017	−4.5	0	20.11.2017	0.06
30.11.2017	−7.2	0.4	30.11.2017	0.03
10.04.2018	0.6	2.7	10.05.2018	0.14
20.04.2018	−0.7	2.8	20.05.2018	0.17
30.04.2018	3.6	1.8	30.05.2018	0.21
10.05.2018	3.5	5.2	06.10.2018	0.24
20.05.2018	6.9	2.2		
30.05.2018	4.3	3.08		
10.06.2018	8.2	9.5		

Four different images—band combinations of Sary-Tash—were made at the same period of time each year, which can be interpreted as land cover classifications. The NDVI map allows for the generation of images that might show active biomass and help to make drought and agriculture production forecasts but it can be identified by additional certain level of analysis. Healthy vegetation may be depicted by a band combination of Landsat 7 ETM+. Thus, it shows dense and healthy plants in green and light green; water bodies are dark (either violet or blue). Bare soil and rocks are represented in white, yellow, orange, brown, and red colors. Visually, it is possible to judge how productive and healthy croplands are. It is clear that in time-series analysis of NDVI values over the last 5 years there is no abnormality—no temporal variations except for the year of 2017. Additional investigations are needed further, particularly using imagery with higher resolution to get more accurate results.

Table 15.2 NDVI values from 1993 to 2003

Date	NDVI	Date	NDVI
10.06.1993	0.25	21.07.1999	0.21
12.07.1993	0.23	06.08.1999	0.40
28.07.1993	0.33	23.09.1999	0.24
13.08.1993	0.38	20.05.2000	0.30
30.09.1993	0.27	21.06.2000	0.0
28.05.1994	0.17	07.07.2000	0.33
13.06.1994	0.14	24.08.2000	0.30
29.06.1994	0.33	07.05.2001	0.14
10.09.1994	0.30	23.05.2001	0.24
15.05.1995	0.18	08.06.2001	0.21
06.09.1996	0.27	26.07.2001	0.37
22.09.1996	0.17	28.09.2001	0.17
09.09.1997	0.21	30.08.2002	0.28
08.06.1998	0.28	16.07.2003	0.32
11.08.1998	0.35	01.08.2003	0.37
12.09.1998	0.19	18.09.2003	0.24
28.09.1998	0.16		

Fig. 15.10 NDVI map of Sary-Tash, April, 2013

Fig. 15.11 NDVI map of Sary-Tash, April, 2014

Fig. 15.12 NDVI map of Sary-Tash, April, 2015

Fig. 15.13 NDVI map of Sary-Tash, April, 2016

Fig. 15.14 NDVI map of Sary-Tash, April, 2017

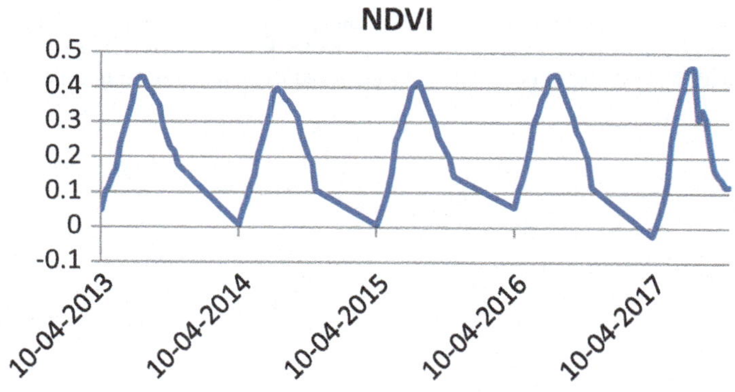

Fig. 15.15 Time-series plot of NDVI over the last 5 years

Acknowledgements This research was accomplished thanks to the financial support of USAID and PEER regional project.

References

1. Arora VK, Boer GJ (2010) Uncertainties in the 20th century carbon budget associated with land use change. Global Change Biol 16:3327–3348
2. Foody GM, Arora MK (1996) Incorporating mixed pixels in the training, allocation and testing stages of supervised classifications. Pattern Recogn Lett 17:1389–1398
3. Mirzabaev A, Goedecke J, Dubovyk O, Djanibekov U, Le Q, Aw-Hassan A (2016) Economics of land degradation in central Asia. In: Nkonya E, Mirzabaev A, von Braun J (eds) Economics of land degradation and improvement—a global assessment for sustainable development. Springer International Publishing, Heidelberg, pp 261–290
4. Kerven C, Steimann B, Ashley L, Dear C, Rahim I (2011) "Pastoralism and farming in central Asia's mountains: a research review." MSRC background paper no. 1. https://www.ucentralasia.org/downloads/pastoralism_and_farming_in_central_asia_mountains
5. Shigaeva J, Wolfgramm B, Dear C (2013) "Sustainable Land Management in Kyrgyzstan and Tajikistan: a research review." MSRI background paper no. 2. https://msri.ucentralasia.org/events.asp?Nid=589
6. Kulikov M, Schickhoff U (2017) Vegetation and climate interaction patterns in Kyrgyzstan: spatial discretization based on time series analysis. Erdkunde 71:143–165. https://doi.org/10.3112/erdkunde.2017.02.04
7. Crewett W (2012) Improving the sustainability of pasture use in Kyrgyzstan. Mt Res Dev 32:267–274. https://doi.org/10.1659/MRD-JOURNAL-D-11-00128.1
8. Warikoo K (ed) (2014) The other Kashmir: society, culture and politics in the Karakoram Himalayas. Copyright © institute for defence studies and analyses, New Delhi. ISBN 978-81-8274-797-5
9. Phiri D, Morgenroth J (2017) Developments in landsat land cover classification methods: a review. [ebook] New Zealand school of forestry, University of Canterbury. Available at: https://www.mdpi.com/2072-4292/9/9/967/pdf [Accessed 13 May 2018]

10. Zhu Z, Shilong P, Ranga BM, Huang M, Zhenzhong Z, Josep GC, Philippe C, Stephen S, Pierre F, Almut A, Chunxiang C, Lei C, Etsushi K, Charles K, Yue L, Xu L, Yongwen L, Ronggao L, Jiafu M, Ning Z (2016) Greening of the earth and its drivers. Nat Clim Change 6

Chapter 16
Investigation of the Various Aspects of the Kafirnigan River Basin, Tajikistan

Z. Z. Nasriddinov, M. Abdusamatov, A. S. Kodirov, J. B. Niyazov, and N. S. Mirakov

Abstract The article is devoted to and carried out in the framework of the PEER NAS USAID project "Integrated Water Resources Management and Strategic Environmental Assessment of Kabul and Amu Darya Rivers". The goal of this article is to assess the environmental and socio-economic situation in the Kafirnigan River basin in face of demographic growth, climate and hydrological regimes changes. In this research, the methods of comparative analysis, field observations, physical and chemical assessments of water quality and GIS technologies are applied. The basis of research and study of these factors is to propose options for managing the water resources of the Kafirnigan River basin and improving the living conditions of the local population. The article pays attention to modern processes of formation and use of water resources on the example of the Kafirnigan River basin of the Republic of Tajikistan. The article discusses the scientific basis of using remote sensing data, in particular, the Normalized Difference Vegetation Index (NDVI), primarily to assess land degradation at various scales and in a wide range of applications, including the stability of agroecosystems. The assessment of environmental and socio-economic of the Kafirnigan River basin shows that the standard of living of the population has different features depending on the zones of formation, dispersion and transit of the flow. Due to the fact that the flow formation zone is located around the city of Dushanbe—the capital of Tajikistan, the majority of the population has greater economic income than other areas of the Kafirnigan River basin.

Keywords Kafirnigan River · Water resources · IWRM · NDVI · Socio-economic assessment

Z. Z. Nasriddinov (✉)
Tajikistan Mining Metallurgical Institute, 6 Moskovskaya Str., Dushanbe, Tajikistan

M. Abdusamatov · A. S. Kodirov · J. B. Niyazov · N. S. Mirakov
Institute of Water Issues, Hydropower and Environment, 14a Aini Str., Dushanbe, Tajikistan

Z. E. Kulenbekov and B. D. Asanov (eds.), *Water Resource Management in Central Asia and Afghanistan*, Springer Water, https://doi.org/10.1007/978-3-030-68337-5_16

16.1 Introduction

In general, Central Asian states have sufficient water resources, but uneven distribution, intensive population growth, increased water use and other factors contribute to water scarcity in many parts of the region. The introduction of water-saving technology (drip, subsurface, sprinkling), the reuse of collector-drainage and wastewater, the use of automated devices for surface irrigation and the improvement of management methods are some of the mechanisms for water saving by increasing the efficiency of irrigation systems [1].

The inefficiency of the existing water management system, the deterioration of the irrigation and drainage system built in the past century, as well as the transition to a market economy and the involvement of water users in water management confirm the need for a transition to Integrated Water Resources Management (IWRM). In this regard, IWRM issues in pilot zones (rivers, target villages) are also relevant, as one of the main tasks of IWRM is to promote the efficient use of water in the interests of sustainable welfare of the society and environmental safety [2].

To address the issues, water resources monitoring, improving the water accounting system, using new devices in determining the physical and chemical properties of soil and water, and determining the rate (flow) of water, both in canals and in rivers, are important. Another important condition is the assessment of the ecological and socio-economic situation (SES) using modern technologies. The purpose of the research is to assess the environmental and socio-economic situation in the settlements of the Kafirnigan River Basin of the Republic of Tajikistan.

Amu Darya and Syr Darya are the largest rivers of Central Asia. The sources of these rivers originate in the mountains of Tajikistan, Kyrgyzstan, Afghanistan and China. The Amu Darya takes its name after the confluence of the Panj and Vakhsh rivers in Tajikistan, the Syr Darya after the confluence of the Kara Darya and Naryn rivers in Kyrgyzstan. The major tributaries of the Amu Darya are the Panj, Vakhsh, Zeravshan and Kafirnigan rivers, whose share in the total water resources of this basin is 82.5% presents data on the formation of surface runoff in the Aral Sea basin [3].

Comparing the data in Table 16.1 based on river flow in Central Asia, it can be seen that water is mainly used for irrigation and other needs in the states located in the middle and lower reaches of the rivers: Kyrgyzstan—4.3%, Tajikistan—11.8%, Uzbekistan—51.5%, Turkmenistan—21.2%, Kazakhstan—9.5% and Afghanistan—1.7% [4].

Regardless of the water availability in each country in Central Asia or a particular river basin, the issues of protection and rational use of water resources are relevant. Thus, IWRM issues in pilot zones (target villages) are also important components for planning purposes. Along with other principles of IWRM (such as participation of women, the public in the process of water resources management, etc.), the introduction of water-saving irrigation technologies and the protection of aquatic ecosystems are very important priorities. In Tajikistan, the Kafirnigan River was chosen as a pilot project, it flows through the territory of 4 districts (Vahdat, Rudaki, Kabadian,

Table 16.1 Surface water resources of the Aral Sea basin

Country	Amu Darya basin (km^3/year)	Syr Darya basin (km^3/year)	Aral Sea basin	
			(km^3/year)	(%)
Kazakhstan	–	4.50	4.50	3.9
Kirgizstan	1.90	27.4	29.30	25.3
Tajikistan	62.9	1.1	64.00	55.4
Turkmenistan	2.78	–	2.78	2.4
Uzbekistan	4.70	4.14	8.84	7.6
Afghanistan	6.18	–	6.18	5.4
Central Asia	78.46	37.14	115.6	100.0

and Shaartuz) of Tajikistan. Vakhdat district is a zone of flow formation, Rudakinsky district is a dispersion zone and Kabadion and Shaartuz districts are transit zones of the Kafirnigan River basin.

16.2 Study Area

The total length of Kafirnigan River is 387 km, the catchment area of the basin is 11.6 km^2. The average discharge of the river is 164 m^3/s; however, depending on the season and weather conditions, the flow varies from 30 to 1200 m^3/s. The glacial region of the Kafirnigan River basin is located in the South of the Tien Shan mountain system (on the southern slopes of the Gissar Range, extending from the eastern part of the Zeravshan Range). The highest altitude is about 5000 m above sea level. The energy potential of the river is estimated at 2883 kW [5].

Hydrologically, the Kafirnigan river basin is relatively well studied. The Kafirnigan River has a snow-glacial type of feed, and high water lasts from March to September with the maximum flow of water in June (headwater) and from February to the end of August with the maximum flow in April downstream. The average annual turbidity of the water in the lower reaches is more than 1500 g/m^3 [5].

In zones located above 3000 m, the climate of the Kafirnigan River basin is characterized by cold summers and moderately severe snowy winters. The Kafirnigan River combines many other large and small rivers and tributaries. This river passes through irrigated land in central and southern Tajikistan before it reaches the Amu Darya River. The Kafirnigan River is very diversely used by local dekhkan (farmer) farms and has a rich historical resource management mechanism that is interesting and relevant to study with a view to developing integrated management and environmental assessment methods to further implement or adjust action plans to the region [5].

16.3 Methodology

During the fieldwork in the Kafirnigan River basin, the newest devices available to the project were used: a Land Mapper ERM-02, which measures electrical resistance and soil conductivity allowing for rapid mapping and monitoring of agricultural land as well as geological, hydrological and environmental studies; an RS485 electromagnetic velocimetry; a water flow meter; weather stations; a PH-meter for the determinant of water acidity and temperature; and a satellite GPS-receiver (Trimble R1) to determine the exact geographic location of the objects of study.

Through the interpretation of satellite imagery, it is possible to analyze many processes occurring on Earth. The presence of a collection of images from different years, months and consideration of the development of phenomena of interest over the long term. One of the predominant objects of the study of Earth from space is plant communities. In particular, great attention is paid to the dynamics of the areas and structure of the wooded territory. Changes occurring in forests can be associated with clear-cutting logging, forest fires, and the loss of forest stands from other negative natural and anthropogenic factors. These types of changes in forest communities can be identified by the data of multi-zone images [6].

A characteristic feature of vegetation and its state is spectral reflectivity, which is characterized by large differences in the reflection of radiation of different wavelengths [7]. Knowledge of the relationship between the structure and the state of vegetation with its spectral-reflective abilities allows for the use of satellite images for mapping and identifying vegetation types. Based on the combination of brightness values in certain channels, informative for highlighting the object under study, and calculating the "spectral index" of an object using these values, an image is constructed corresponding to the index value in each pixel. This allows for the selection of the object under study and evaluation of its state [8]. Spectral indices used to study and assess the state of vegetation have the generally accepted name of vegetation indices and allow for the automation of the process of identifying various plant communities and assessing their qualitative characteristics.

16.4 Results and Discussion

The RK900-05 Weather Station device installed by the research team in the Kafirnigan River basin (RIKA-Wireless Home Weather station) measured 12 parameters (precipitation, evaporation, humidity, wind speed, lighting, etc.) and showed reliability in work. A comparison of the data from the RK900-05 with the data from the station of the Agency for Hydrometeorology of Tajikistan (Esanbay station) for 2018 showed that they were very close. Figure 16.1 shows the average monthly temperature and precipitation data in 2017 obtained from the RK900-05 Weather Station.

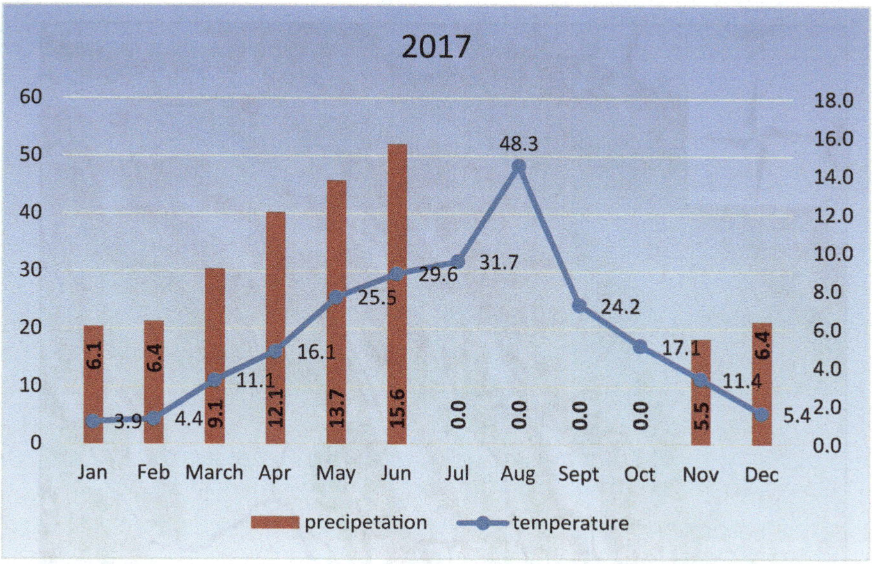

Fig. 16.1 The average monthly temperature and average monthly rainfall for 2017 precipitation (mm); temperature (°C)

The calculation of most of the vegetation indices is based on the two most stable sections of the spectral reflectivity curve of plants and are not dependent on other factors. The red zone of the spectrum (0.62–0.75 μm) accounts for the maximum absorption of solar radiation by chlorophyll, and the near-infrared zone (0.75–1.3 μm) accounts for the maximum reflection of energy by the leaf's cellular structure (Fig. 16.2).

Thus, high photosynthetic activity is associated, as a rule, with a large phytomass of vegetation, and leads to lower values of reflection coefficients in the red region of the spectrum and large values in the near infrared. The relationship of each other allows researchers to clearly separate the vegetation from other natural objects [9]. One of the most well-known vegetative indices was chosen for this study—the Normalized Difference Vegetation Index (NDVI).

To identify objects on the Earth's surface, it is useful to combine different channels to obtain images in natural colors (if red, green, and blue channels are used) or pseudo-color images (if another combination of channels is used). For example, with LANDSAT ETM +, the combination of channel numbers 3-2-1 gives a realistic image in natural colors. However, pseudo-color images are more suitable for many purposes. For example, channel number combinations, such as 4-3-2, 7-4-2, or 4-5-3, are widely used to determine vegetation or water surfaces, respectively [10].

Semiautomatic extraction of objects of interest (vegetation, water mirrors, etc.) includes the construction of a ratio of different channels. The construction of the relationship eliminates the relationship between the ranges, thus eliminating unnecessary data and retaining very specific information. This approach is suitable if it is

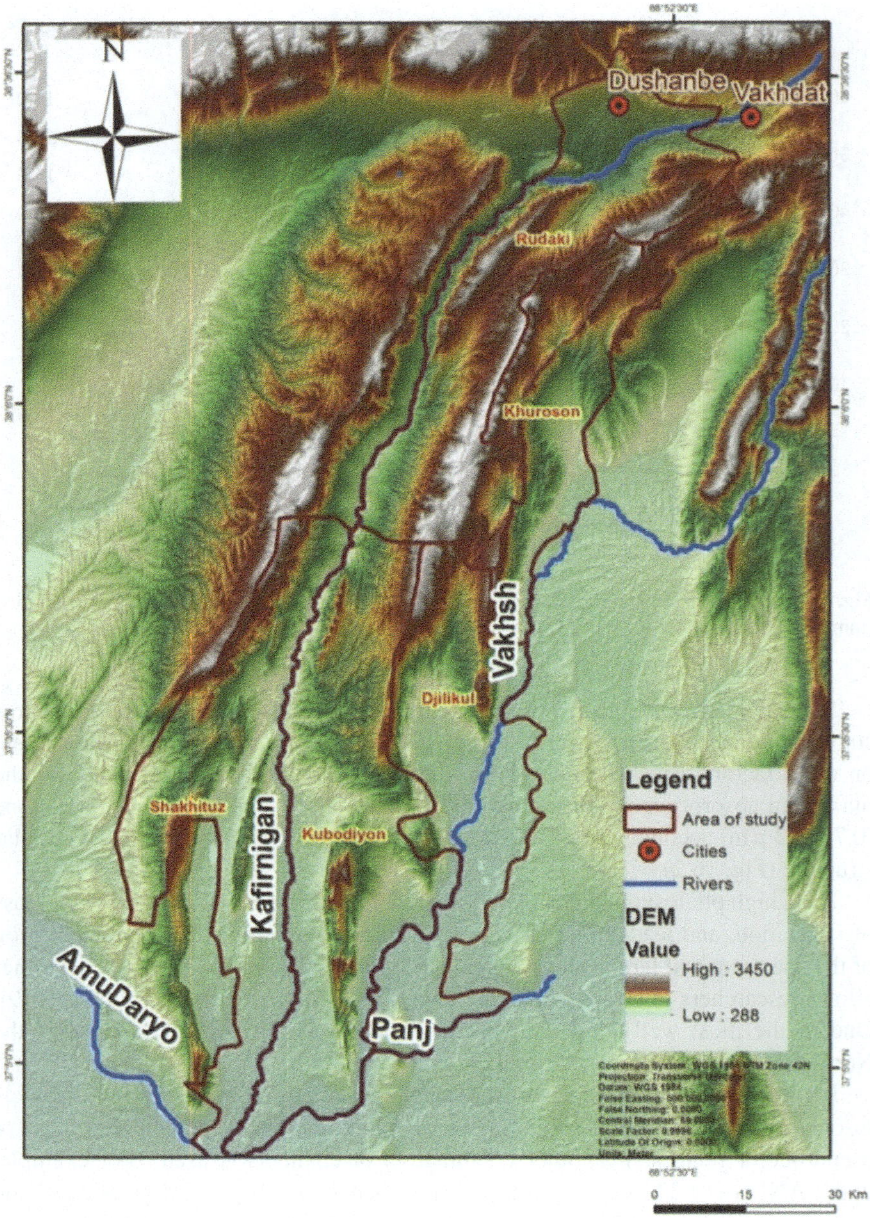

Fig. 16.2 The research area of the Kafirnigan River basin

necessary to select a specific type of object which is strongly reflected in one channel and weakly in another.

Allocation of vegetation-covered areas is widely distributed by calculating the normalized differential vegetation index NDVI:

$$\mathrm{NDVI} = \frac{\mathrm{NIR} - \mathrm{RED}}{\mathrm{NIR} + \mathrm{RED}},$$

where NIR is the reflection in the near-infrared region of the spectrum, RED—reflected in the red region of the spectrum.

According to this formula, the density of vegetation (NDVI) at a certain point in the image is equal to the difference of the intensities of the reflected light in the red and infrared ranges, divided by the sum of their intensities.

The range of NDVI values varies from -1 to 1, in which the highest values indicate the presence of lush vegetation and low values indicate rare vegetation cover. NDVI values should be classified in order to clearly identify vegetation areas. It is necessary to determine equal values for the classified visualization of NDVI separately for each satellite scene in order to calculate the difference between the scenes (Fig. 16.3).

The use of automatic selection indices includes some explanations: in addition to determining the actual vegetation-covered zones, the results of automatic selection using NDVI will always contain two types of errors: (a) zones are defined as vegetation-covered when they are not exactly those, and vice versa: (b) zones are defined as not covered with vegetation, when in fact they are clearly covered with vegetation. To calculate these errors, an accuracy estimate can be performed using ground survey data or images obtained through visual interpretation.

During field studies, a survey was conducted among residents of the Vahdat (Tangai and Ramit village), Rudaki (Sarikishti village) and Kabadiyan (Tartki village) districts, according to a pre-compiled questionnaire that covered different socio-economic situations (SES). One of the indicators was the activity of the Water Users Association (WUA) in the studied areas. According to the information which was received from the Agency for Land Reclamation and Irrigation of Lands under the Government of the Republic of Tajikistan, as of September 1, 2017, there are 383 WUAs operating in the republic which serve 387,559 ha of irrigated land. Of this number, in the studied area of the Kafirnigan River basin, only 45 WUAs serving 52,176 ha were in operation. On average, 1 WUA accounts for 1159 ha of irrigated land in the Kafirnigan River basin.

WUA efficiency is still low. They do not have trained specialists and they are not provided with transport and earthmoving equipment. Not all WUAs have a material base, including offices. Another important indicator of SES was the family cattle breeding of the studied basin (Table 16.2).

An analysis of family cattle breeding shows that, in the areas of runoff formation and the middle flow of the Kafirnigan River (Vahdat and Rudaki districts), the population has less livestock (about 20% of the families) than in the area of the dispersion (irrigation) districts of Kabadiyan, where are more than 90% of residents have livestock.

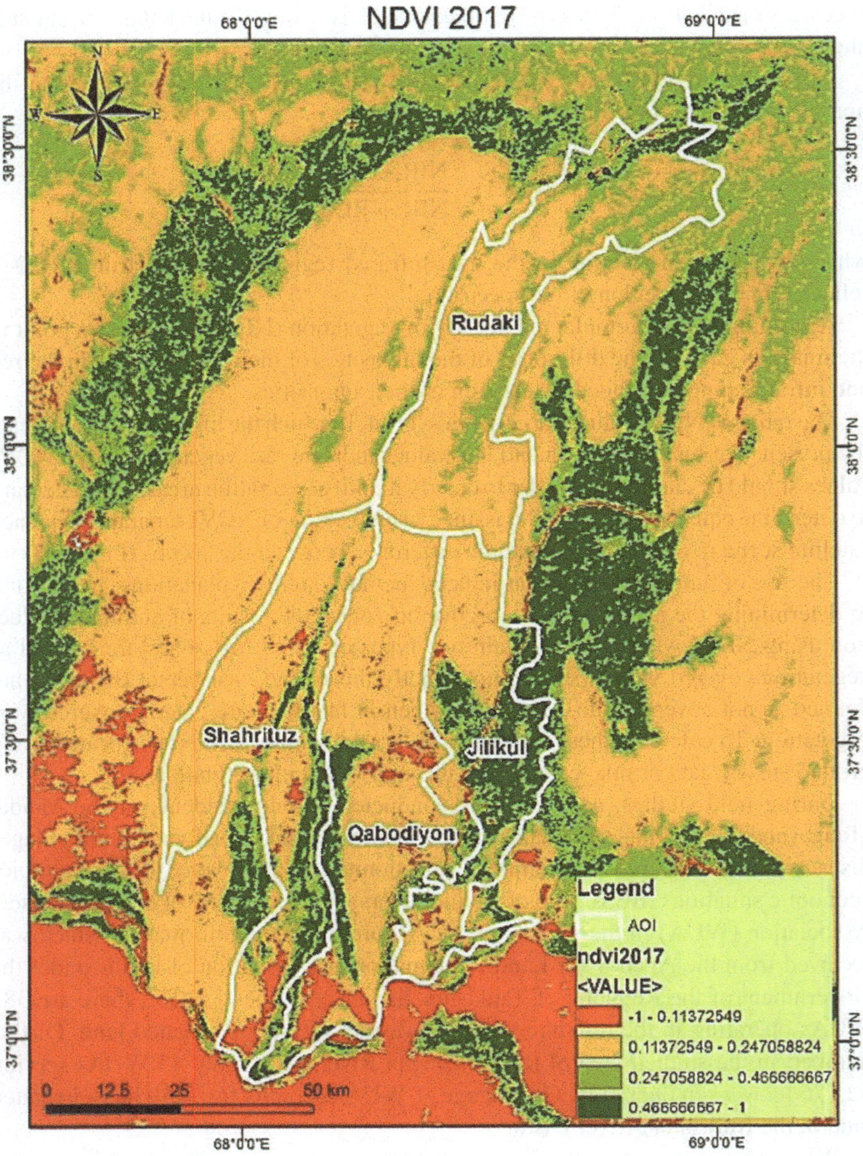

Fig. 16.3 NDVI for the Kafirnigan River basin, 2017

The water quality of the Kafirnigan River mainly complies with the norms of maximum permissible concentration (MPC). In the flood period, there is an excess of MPC for suspended substances. Water samples for chemical analysis were taken from three points (upper reaches, middle part, and dispersion zone) in 2018, and

Table 16.2 Families involved in cattle breeding (%)

Polling place	Cattle	Sheep	Goat	Horses	Chickens	Their absence (%)
Vahdat, Tangai and Ramit villages	5	5	5	0	5	80
Rudaki, Sarikishti village	5	5	5	0	5	80
Kabodiyon, Tartki village	20	10	10	2	50	8

these showed that the pH was within the normal range (6.85–7.21), as was TDS (300–600 mg/l). Of the heavy metals in water, the presence of molybdenum (Mo), cadmium (Cd), nickel (Ni), chromium (Cr), and others that did not exceed the permissible rate was noted.

16.5 Conclusion

Studies conducted in the Kafirnigan River basin have shown the need for introducing water-saving technologies, integrated water resources management, improving the ecological status of rivers, socio-economic conditions of the population, improving the efficiency of irrigated land, planning and introducing efficient use of water resources of the partner countries. And the use of GIS technology allows you to determine the vegetation index (NDVI), which reflects the total amount of vegetation and is used to assess its condition and build environmental-climatic maps.

Acknowledgements This research was accomplished thanks to the financial support of the USAID and PEER regional project.

References

1. Kodirov AS (2015) The role of water users associations in improving water management (on the example of the Gissar Valley). In: Proceeding of dedicated to the international decade for action “water for life, 2005–2015”. TajikNIIGiM, Dushanbe, pp 142–147
2. Tushaar Sh, Barbara VK (2006) Is India ripe for integrated water resources management? Fitting water policy to national development context. Econ Polit Week 41(31):3413–3421
3. Nuraliev K, Abdusamatov M, Latipov RB (2011) Water resources of Tajikistan: initiatives, strategy and perspectives. Asia-Print Publishing House, Dushanbe, 219 p
4. Kodirov AS (2016) Formation and dynamics of the hydrological regime of the Vakhsh river basin of the Republic of Tajikistan. Diss Cand Tech Sciences. Kodirov A.S., Dushanbe-Bishkek, 24 p
5. Catalog of glaciers SSRR, vol 14, Central Asia, Issue 3, Part 5, L. Hydrometeoizdat (1980)
6. Terekhin EA (2013) The method of mapping perennial changes in forests based on the analysis of their spectral characteristics over the series of satellite data of different times. In: Study of earth from space, № 5, pp 62–69

7. Antonov VN, Sladkikh LA (2009) Monitoring the state of crops and forecasting the yield of spring wheat according to the ERS data. Geomatics № 3:50–53
8. Cherepanov AS, Druzhinina EG (2009) Spectral properties of vegetation and vegetation indices. Geomatics № 3, WITH 28–32
9. Crippen RE (1990) Calculating the vegetation index faster. Remote Sens Environ 34:71–73. Bartalev SA (2005) Methods of using time series of satellite images of high spatial resolution for assessing the extent and dynamics of deforestation of taiga forests. Bartalev SA, Kuryatnikova TS, Stibig KhYu. Modern problems of remote sensing of the earth from space. T. 2. № 2 WITH 217–227
10. Landsat Data Continuity Mission. Press Kit (2013) [Electronic resource]. URL: www.nasa.gov/pdf/723395main_LDCMpresskit2013-final.pdf

Chapter 17
New Approaches and Advanced Methodology in Integrated Water Resources Management: Amu Darya River Basin

Zheenbek E. Kulenbekov, E. Khazieva, Sagynbek Zh. Orunbaev, and Baktyiar D. Asanov

Abstract Agriculture and livestock are the main sectors of a river basin region's economy, particularly along the Amu Darya River. This makes sustainable pasture and land management critical for human well-being, economic stability, social welfare, and ecosystem resilience. Both human-induced and natural factors play a key role in the sustainability issues of rural mountainous communities in the Amu Darya river basin that rely heavily on land resources. This study focuses predominantly on finding a linear relationship between values of Normalized Difference Vegetation Index (NDVI) and climatic variables such as air temperature, land surface temperature, precipitation using new approaches and advanced methodology in the Google Earth Engine (GEE). This helps gain an understanding of the seasonal and inter-annual behavior and dynamics of the vegetative characteristics. Overall, the implications of this study are directed toward the general understanding of the interaction between terrestrial ecosystems and climate change. The study encompassed three river basins along the Amu Darya-Kunduz River basin in Afghanistan, the Kafirnigan River basin in Tajikistan, and the Kyzyl-Suu River basin in Kyrgyzstan.

Keywords NDVI · GEE · River basin · Air temperature

17.1 Introduction

17.1.1 A Subsection Sample

While water resource conservation, management, and sustainable development are the integral part of economic development and environmental restoration in river basins along the Amu Darya, there has been no comprehensive study conducted on the

Z. E. Kulenbekov (✉) · S. Zh. Orunbaev · B. D. Asanov
American University of Central Asia, 7/6 Aaly Tokombaev Str., Bishkek, Kyrgyzstan
e-mail: kulenbekov_z@auca.kg

E. Khazieva
Moscow State University, 1 Kolmogorova Str., Moscow, Russia

Z. E. Kulenbekov and B. D. Asanov (eds.), *Water Resource Management in Central Asia and Afghanistan*, Springer Water, https://doi.org/10.1007/978-3-030-68337-5_17

impacts of climate change, the IWRM, land cover, and environmental degradation. The lack of study on the impact of climate change, hydrology and water quality data is one of the main challenges for sustainable water resources management. Land cover information plays a significant role in climate change studies as well as in understanding complex interconnectedness between human activities and global change. Accurate and up-to-date land cover information also plays a critical role in sustainable resource management, planning, and monitoring activities.

Conventional ways of land cover classification, generally constrained by field surveys, are insufficient and time-consuming. In addition, despite the fact that remote sensing has long served as a generator of more accurate land cover information, there are still problems that occur while working with the extraction of land cover information. Improper and mishandled satellite image treatments may result in inappropriate output when classifying land cover. Much has been recently researched on climate change and vegetation response to identifying spatio-temporal variations in Central Asia. Earth's soil has always been subject to change and alteration due to both natural and anthropogenic factors.

17.1.2 Satellite Data

Satellite technologies and new ways of data analysis have started playing an important role in agricultural monitoring systems. Until recently this area of knowledge was only available to limited groups of remote sensing and geospatial practitioners. However, the latest computer achievements made the benefits and technology of satellite inquiry manageable for non-experts and decision-makers. Until recently, it was highly time-consuming to gather Landsat or Sentinel images, pre-process them and analyze them. Not only did it require high-performance personal computers, but also a wide internet bandwidth. However, with the advent of the new cloud approach, the process of data collection and analysis has been modified and simplified. After the Landsat data was made publicly available in 2008, Google started to collect this huge data set on their cloud platform. The launch of Sentinel-1A, Sentinel-1B in 2014 and 2016, respectively, [14] Sentinel-2 in 2015, and Sentinel-3 in 2018 has provided an opportunity to work with high resolution (up to 5 m) images and high frequency of observations (3–5 days). After Google developed a cloud platform, the development of the Google Earth Engine (GEE) now allows searching for different data and running a variety of geospatial analyses on Google's IT infrastructure [7]. GEE has a data catalog with freely available geospatial data, including information from the above-mentioned sensors. The GEE data catalog consists of different free geospatial data sets, including satellite imageries (optical and radar), information about climate (temperature, precipitation, etc.), relief (digital elevation models), land cover maps, and a variety of socioeconomic variables [7].

The GEE data catalog has more than 240 public data sets, collecting about 4000 new images every day (Table 17.1).

Table 17.1 Following datasets from GEE might be used for vegetation and water surface analysis

Data source	Data availability (time)	Description	Collection ID in GEE
Landsat 1 MSS	1972–1978	4 bands; spatial resolution 30–60 m	LANDSAT/LM01/C01/T1
Landsat 2 MSS	1975–1982	4 bands; spatial resolution 30–60 m	LANDSAT/LM02/C01/T1
Landsat 3 MSS	1978–1983	4 bands; spatial resolution 30–60 m	LANDSAT/LM03/C01/T1
Landsat 4 ETM	1982–1993	Atmospherically corrected surface reflectance; 4 bands; spatial resolution 30–60 m	LANDSAT/LT04/C01/T1_SR
Landsat 5 ETM	1984–2012	Atmospherically corrected surface reflectance; 4 bands; spatial resolution 30–60 m	LANDSAT/LT05/C01/T1_SR
Landsat 7	1999–2018	Data calibrated top-of-atmosphere (TOA) reflectance; 8 bands; spatial resolution 15–30 m	LANDSAT/LE07/C01/T1_RT_TOA
Landsat 8	2013–2018	Data calibrated top-of-atmosphere (TOA) reflectance; 11 bands; spatial resolution 15–30 m	LANDSAT/LC08/C01/T1_RT_TOAmport
Sentinel-1 SAR	2014–2018	Provides data from a dual-polarization *C-band Synthetic Aperture Radar* (SAR); 4 polarization combinations (single VV or HH, and dual band VV + VH and HH + HV)	COPERNICUS/S1_GRD
Sentinel-2 MSI	2015–2018	16 bands; spatial resolution 10–60 m	COPERNICUS/S2
Sentinel-3 OLCI	2016–2018	24 bands; spatial resolution 300 m	COPERNICUS/S3/OLCI
MODIS land cover type	2000–2003	Provides data characterizing five global land cover classification systems (IGBP, UMD, LAI/fPAR, NPP, PFT schemes)	MODIS/051/MCD12Q1

(continued)

Table 17.1 (continued)

Data source	Data availability (time)	Description	Collection ID in GEE
GFSAD1000	2000–2001	Provides the spatial distribution of the five major global cropland types (wheat, rice, corn, barley, and soybeans)	USGS/GFSAD1000_V0
Hansen global forest change	2000–2017	Provides a time series analysis of Landsat images in characterizing global forest cover and its change	UMD/hansen/global_forest_change_2017_v1_5
Global PALSAR forest/non-forest map	2007–2018	Is generated by classifying the PALSAR-2/PALSAR SAR image; spatial resolution 25 m	JAXA/ALOS/PALSAR/YEARLY/FNF
GIMMS NDVI	1981–2013	Vegetation index that was generated from several NOAA's AVHRR sensors; spatial resolution 5 arc minutes	NASA/GIMMS/3GV0
MODIS	2000–2019	2 bands (NDVI, EVI), spatial resolution 250 m	MODIS/006/MOD13Q1
Water occurrence change intensity	1984–2015	JRC global surface water mapping layers. The location and temporal distribution of surface water from 1984 to 2015 JRC global surface water mapping layers. The location and temporal distribution of surface water from 1984 to 2015	JRC/GSW1_0/GlobalSurfaceWater
Water occurrence			JRC/GSW1_0/GlobalSurfaceWater

Due to good infrastructure and flexible tools, GEE can support efficient data processing used in different studies, such as long-term water surface monitoring [13], flood, and drought monitoring [15], crop condition assessment [11, 16, 20], and land cover changes analysis [17]. The NDVI time series analysis has become a useful research tool that allows for understanding the changes to vegetation cover over time. When using GEE, it is possible to analyze all Landsat images beginning in 1972 until the present day for a particular area. However, it does require additional processing such as pan-sharpening, mosaicking, cloud masking, etc. Also, there are some already processed data available, such as the MODIS NDVI data sets.

17.1.3 Case Studies

Most hydrological and climate data collection in Afghanistan was interrupted in the early 1980s as a consequence of war, only partially resuming after 2004 [12]. The Kunduz River Basin is a tributary of the Amu Darya River in North Afghanistan and the Panjshir-Ghorband River Basin a tributary of the Kabul River Basin.

In Tajikistan, the Kafirnigan River was chosen as a focus area, it flows through the territories of 4 districts (Vahdat, Rudaki, Kabadian, and Shaartuz) in Tajikistan. Vakhdat district is the zone of flow formation, Rudakinsky district is the dispersion zone, and the Kabadion and Shaartuz districts are transit zones of the Kafirnigan River Basin.

Livestock husbandry and agriculture are both essential activities in the Kyzyl-Suu River Basin of the Chon-Alay region in Kyrgyzstan and play critical roles in the local economy. Animal husbandry is highly dependent on climate conditions, as the climate has a great impact on pastureland and fodder. Periodically, natural disasters such as landslides and earthquakes, as well as heavy precipitation and long winters, severely damages local economies.

17.2 Study Area

17.2.1 Topography and Climatic Condition

The Panjshir-Ghorband River originates from the south and the Kunduz River from the north side of the Hindukush Mountains and flows through Northern Afghanistan, finally reaching the Amu Darya. The Panjshir-Ghorband watershed encompasses about 12,963 km^2, and the Kunduz watershed covers 28,023 km^2 in area [5]. Both the Panjshir-Ghorband and Kunduz River Basins originate from the Hindukush Mountains. Figure 17.1 shows the hydrology and geographic locations of the study area.

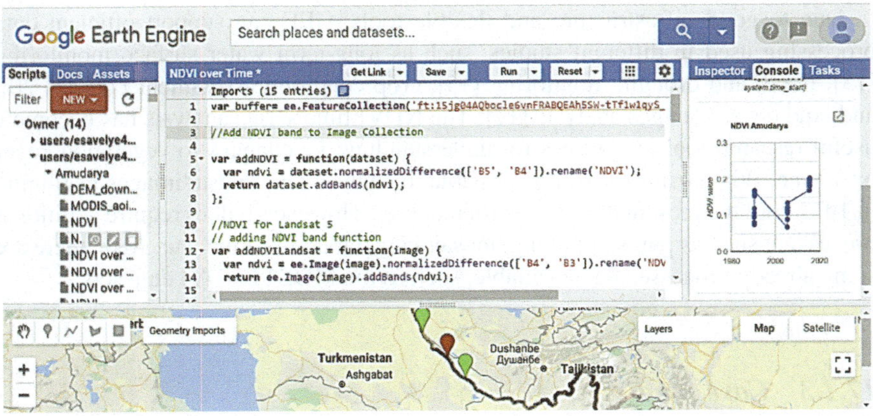

Fig. 17.1 GEE Code Editor interface

As the relief of the River Basins consists of high-altitude mountainous areas ending with lowland areas, precipitation and temperature have high variation within the basins. The annual average precipitation and temperature in the Jabalusarag area, located in the middle of the Panjshir-Ghorband River Basin, is 470 mm and 14.4 °C for 2018. The average annual precipitation and temperature in the Kunduz area is 325 mm and 16.8 °C respectively [1].

Both of the river basins consist of a number of tributaries. The hydrology of the basins is mainly controlled by the high mountains of the Hindukush [6]. Upstream, channels are generally narrow and flow throughout the whole year. Runoff regimes are largely controlled by snow-melt and the high discharge that happens from April to June [10]. Precipitation in these basins mainly occurs in the form of rain, drizzle, snowfall, and hail, and is high during the winter months. The areas near the rivers and the floodplains consist of highly fertile soil with productive agricultural land. This area also comprises the main economic center of the basins. The higher altitude areas in the basins are used partially for rainfed agriculture, but mostly consist of deforested land [19].

The total length of the Kafirnigan River is 387 km, the catchment area of the basin is 11.6 km^2. The average discharge of the river is 164 m^3/s; however, depending on the season and weather conditions, the flow varies from 30 to 1200 m^3/s. The glacial region of the Kafirnigan River Basin is located in the South of the Tien Shan mountain system (on the southern slopes of the Gissar Range, extending from the eastern part of the Zeravshan Range). The highest altitude is about 5000 m above sea level. The energy potential of the river is estimated at 2883 kW [2].

Hydrologically, the Kafirnigan River basin is relatively well studied. The Kafirnigan River has a snow-glacial type of feed, and high water lasts from March to September with the maximum flow of water in June (headwater) and from February to the end of August with the maximum flow in April downstream. The average annual turbidity of the water in the lower reaches is more than 1500 g/m^3 [2].

In zones located above 3000 m, the climate of the Kafirnigan River Basin is characterized by cold summers and moderate to severe snowy winters. The Kafirnigan River combines many other large and small rivers and tributaries. This river passes through irrigated land in central and southern Tajikistan before it reaches the Amu Darya River. The Kafirnigan River is used very diversely by local dehkan (farmer) farms and has a rich historical resource management mechanism that is interesting and relevant to study with the view of developing integrated management and environmental assessment methods to further implement or adjust action plans for the region [2].

The study area covers Chon-Alay and Alay districts of Kyrgyzstan. An administrative center of Chon-Alay district is Daroot-Korgon village and in the Alay district, Sary-Tash village is located. Both Daroot-Korgon and Sary-Tash are located along the river, Kyzyl-Suu. The Daroot-Korgon is situated on the western part of Alay Valley, lying between the Alay and Pamir-Alay mountain ranges.

The climate of the study area is sharply continental and is subject to annual variations in temperature, severely long and cold winters and moderately warm summers. The hottest months are July and August with mean temperatures 15.5 °C and 16.3 °C, respectively.

The influence of the Kyzyl-Suu River on the local neighborhoods is unquestionable. Water sufficiency in Alay Valley was confirmed by surveys in the form of a questionnaire. According to this data, Kyzyl-Suu and its inflow rivers significantly affect the positive development of villages in the Alay Valley. Despite cold climate in the Alay Valley, most people use tap water (35%), groundwater (30%) and the river (25%) as a pure source of drinking water, which covers the needs of the population [2].

17.3 Methods and Materials

Google Earth Engine (GEE) is a cloud-based platform for geospatial analysis in a variety of high-impact fields such as deforestation, disaster and water management, food security, and climate monitoring. GEE revolutionizes remote-sensing data analysis as it provides access to the large data catalog and allows the use of cloud technology for processing. This approach saves resources on the client's side and empowers remote-sensing scientists who need to use large-scale commodity computing resources.

New possibilities in data analysis allow for the gathering and processing of a lot of remote-sensing data for relatively huge areas. Not only does GEE provide cloud computing resources and data sets, but it also includes an agile tool for data searching, loading, analysis, and visualization—Code Editor (Fig. 17.1).

Code Editor is a web-based integrated development environment for GEE API. It uses JavaScript for data analysis and visualization. Users can import their own data and use it along with the data from the GEE catalog. It also provides more than 300 methods and algorithms that can be used for data analysis.

17.3.1 NDVI Trend Analysis (2000–2015)

For long-term analysis, a vegetation index called the Normalized Difference Vegetation Index (NDVI) was used. Usually, NDVI takes the value from +1.0 to −1.0. Bare surface or sand typically is characterized by low values (0.1 or less). Scarce vegetation such as shrubs and bushes shows values from about 0.2 to 0.5. Dense and healthy vegetation often result in high values, more than 0.6–0.7 [18]. The calculation is based on the ability of healthy vegetation to absorb the most visible part of the spectrum and reflect a large amount of the near-infrared lights. Whereas the reflection of unhealthy or sparse vegetation mostly consists of visible lights, the reflection of healthy vegetation is affected most by wavelengths in the near-infrared part of the spectrum. Mathematically, this index is a ratio of spectral reflectance values of red band (Red) and near-infrared (NIR).

The one objective of the research was the analysis of NDVI changes within a case study area from 2000 to 2015. It is necessary to have data from the same satellite in order to accomplish this task. We expected to use Landsat data for these purposes, but the analysis of Landsat 5 showed that there are no images in the Landsat 5 collection covering our case study area (Fig. 17.2).

Landsat 7 collection was also excluded since the images available were characterized by a high level of cloudiness (more than 40%). Sentinel-2 is a perfect solution for real-time analysis because of its relatively good spatial resolution (10 m) and high frequency of observation, it provides imageries for the entire globe every 5–10 days [4]. However, due to the data being available only since 2015, this dataset couldn't be used in our research.

As a result, the MODIS data was chosen for NDVI trend analysis. There are few datasets in GEE catalog that included NDVI values: MODIS Terra Vegetation Indices (GEE ID "MODIS/006/MOD13Q1") with the 250 m spatial resolution and MODIS Terra Daily NDVI (GEE ID "MODIS/MOD09GA_006_NDVI") with the

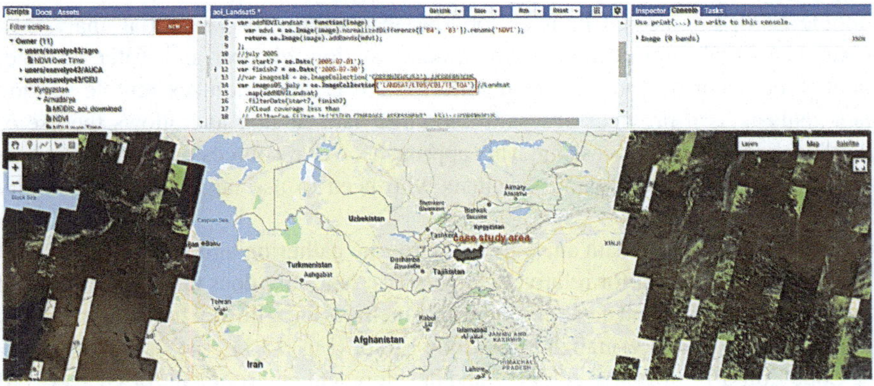

Fig. 17.2 Landsat 5 collection

500 m spatial resolution. The Google team developed the last data set and it was generated from the MODIS/006/MOD09GA surface reflectance composites.

In this research, we used 250 m resolution original MODIS datasets. This product provides information about two vegetation indices: the Normalized Difference Vegetation Index (NDVI) and the Enhanced Vegetation Index (EVI) that minimizes canopy background variations. The EVI index is similar to the NDVI, however due to blue bands in the calculation it manages to avoid distortions from aerosols and other atmospheric particles. The indices were computed using surface reflectance that has been masked by water, clouds, heavy aerosols, and cloud shadows [8].

Calculation showed that summer is the most representative time of the year for this area, we calculated mean values from the NDVI for July of each year. Since the MODIS data is available every 16 days, we analyzed 30 images and calculated the mean NDVI values for the month of July from 2000 to 2015 (Fig. 17.3).

The NDVI values range from -1.0 to $+1.0$ and we had to apply a scale of 0.0001 as it is indicated in the description of the product [3].

The calculated NDVI data was compared with the mean July temperatures. The temperature data set was developed within the NCEP/NCAR Re-analysis Project and provides the surface air temperature every 6 h [9]. Since the product provided

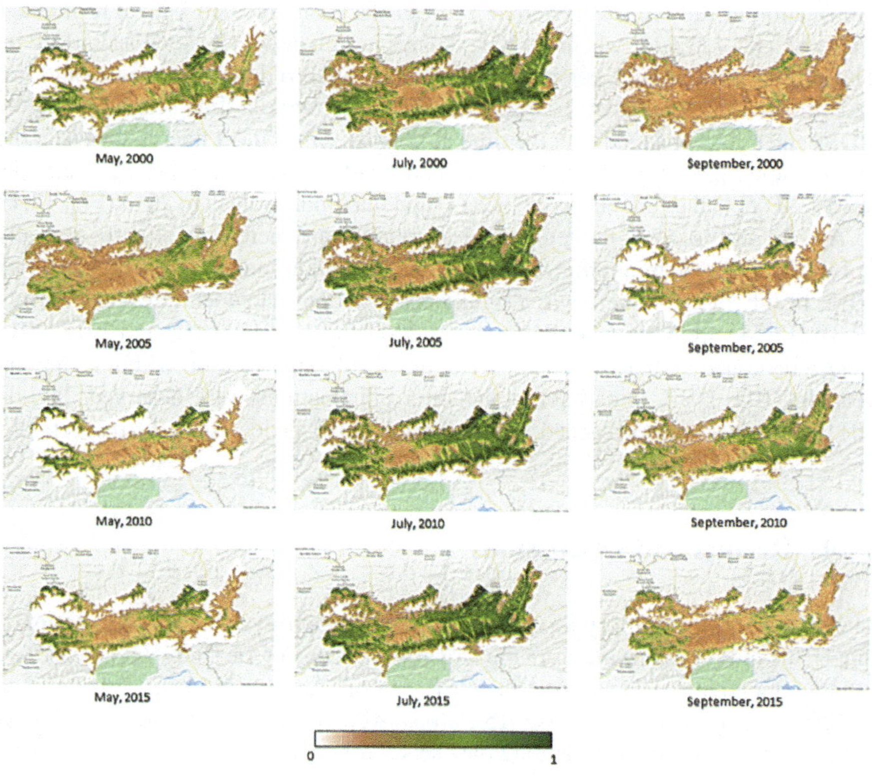

Fig. 17.3 Mean NDVI values for May, July, and September of 2000, 2005, 2010, and 2015

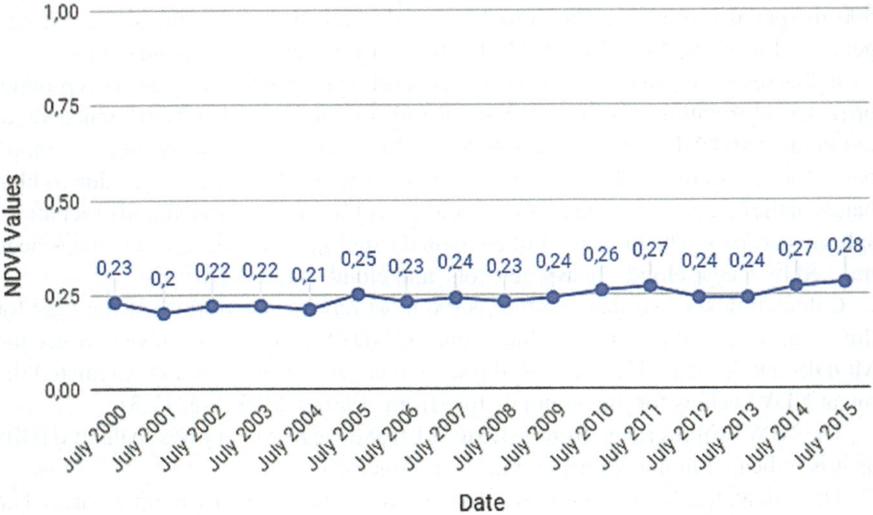

Fig. 17.4 Mean NDVI value for July 2000–2015

information in Kelvin, we converted it to Celsius using the well-known formula: T (°C) = T (K)—273.15, where T (°C)—values in Celsius, T (K)—values in Kelvin. Thus, 1860 (124 images per month) have been analyzed and the mean value for each month was calculated. In Fig. 17.5, we can observe a time series plot of the NDVI and how it changed over the time. It illustrates the NDVI fluctuations over the observational period of 2000–2015. We can vividly observe a trend: as a rule, the lowest NDVI values are observed in 2000 and maximum values in 2015 (Fig. 17.4).

As seen in Fig. 17.5, the temperature variations are increasing slightly annually in July. In 2000, the minimum value was 7.97 and maximum value reached 15.24 by 2015. In July of 2014, there was a significant variation of temperature shifting from 12.3 to 15.24 within one year. So, the years 2000–2001, 2011–2012, and 2014–2015 clearly demonstrate that temperature can drastically change even within a singular year. In 2011–2012, remarkable temperature fluctuations were observed. In July 2011, the highest values reached 12.0 and in July 2010 the lowest 10.39, respectively.

17.4 Results and Discussions

The coverage of the territory was attributed to the availability of one weather station—that is, the more stations there are the larger is the scope of the study. In our case, we used a temperature data set from the GEE catalog. As seen in Figs. 17.6, 17.7 and 17.8, for the time range (2000–2015), a positive correlation coefficient was found between the NDVI and temperature data sets (GEE catalog).

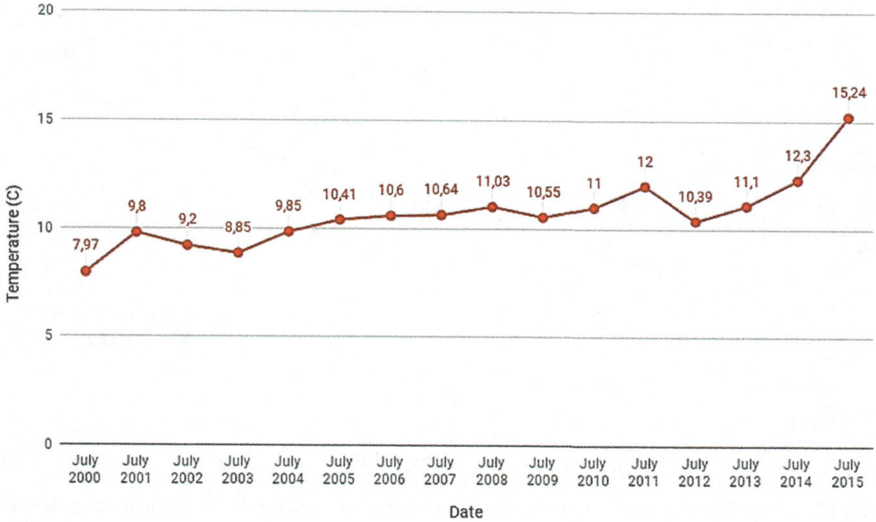

Fig. 17.5 Mean July temperature for 2000–2015

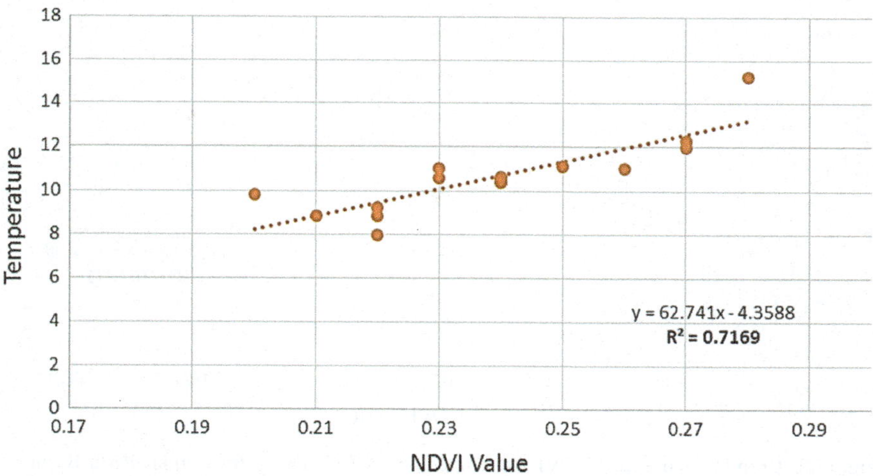

Fig. 17.6 Correlation between NDVI and temperature (GEE catalog) for Kyzyl-Suu River Basin

The NDVI trend analysis showed a gradual increase within 2000–2015, starting from 0.23 in 2000 and the ending value being 0.28 in 2015. Since the NDVI values correspond with the vegetation health and biomass, this growth could imply the expansion of shrubs. Temperature values are increasing as well within the observational period, showing a minimum value of 7.9 in 2000 and a maximum value of 15.2 in 2015.

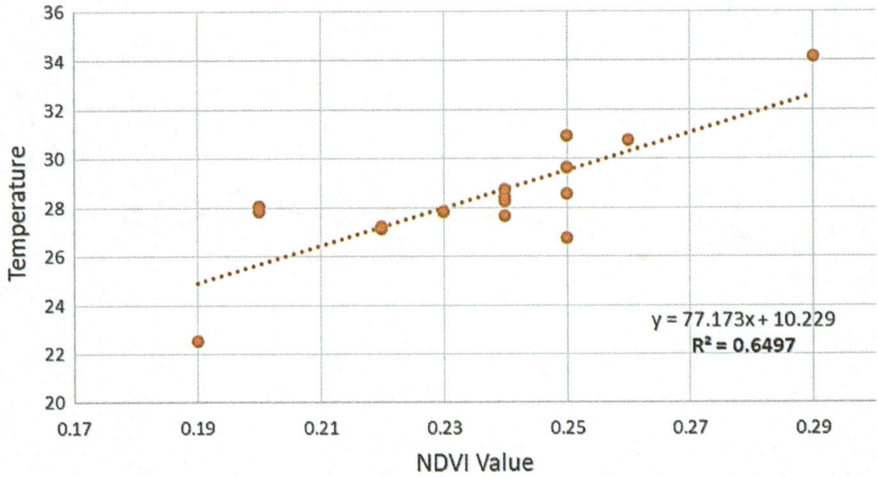

Fig. 17.7 Correlation between NDVI and temperature (GEE catalog) for Kafirnigan River Basin

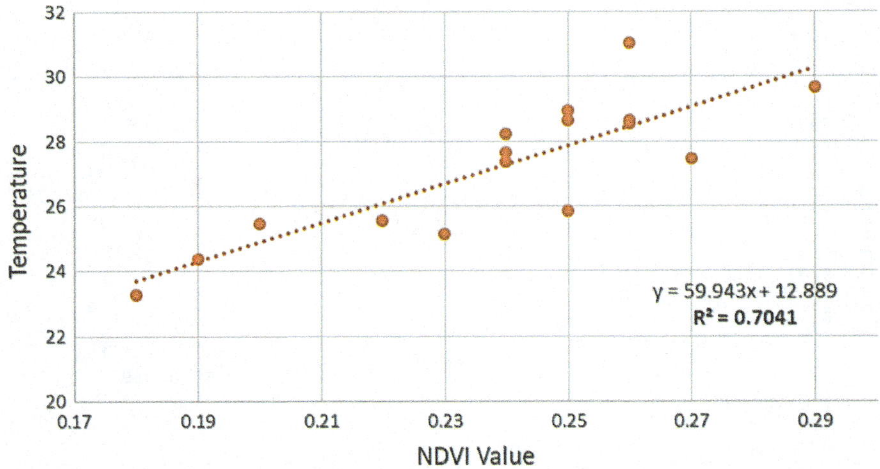

Fig. 17.8 Correlation between NDVI and temperature (GEE catalog) for Kunduz River Basin

For correlation calculations, the NDVI values were compared with temperature information gathered from the field and with the air surface temperature data set from the GEE catalog. In the case of the Kyzyl-Suu River Basin, the correlation between the NDVI and temperature data from the GEE catalog is a strong positive linear relationship—0.71 (Fig. 17.6). However, all results show a good correlation between these two parameters.

In Fig. 17.7, an inter-annual sequence of the NDVI and temperature values in July in the Kafirnigan River Basin. Judging by visual assessment as well as quantitatively, it is clear that a strong positive linear relationship, when using the Pearson correlation—0.81. However, neither graphically nor numerically are we able to identify very strong variations in the NDVI in July of each year. So, based on a 15-year analysis, July is the month with the highest NDVI values with a positive trend: values do not decrease sharply—they are evenly increasing.

Figure 17.8 illustrates the correlation in the Kunduz River Basin and shows the strong positive linear relationship (0.84) between temperature and the NDVI values. Visually, from the figure, it is possible to judge how productive and healthy croplands are in the Kunduz River Basin. An increasing temperature increases the NDVI value. Thus, what is observed is a clear correlation between NDVI and temperature.

17.5 Conclusions

The study encompasses two main goals: finding a relationship between the NDVI and a climatic variable, such as land surface temperature in Kyzyl-Suu, Kafirnigan, and Kunduz River Basins and around villages. The study focused on the one timeframe, 2000–2015. Based on the data available, the first research question was answered: all three river basin areas were found to have positive correlations between vegetation dynamics and climatic parameters. In other parameters, again due to partial data availability of precipitation, no significant interrelationship between the two variables was found. However, the research study gave an answer to whether it is possible to predict one variable knowing the other if there is a linear interrelationship between the NDVI and temperature. If there are only few years with positive linear interdependence within 16 years of observation, it is difficult to claim that there is a significant trend in precipitation. This gives a reason to believe that in the case where the data was in a good state, the analysis would be much more fruitful and reliable.

However, analysis of the NDVI trends over the last fifteen years in the parts of the Amu Darya River Basin clearly illustrates a trend: a strong uphill (positive) linear relationship is definitely a trend.

This study was undertaken partly as a gap of knowledge: the region along the Amu Darya and specifically river basin villages have not been studied sufficiently. There was not enough instrumental data due to the political situation in the region. Potentially, using research questions posed in this paper, future researchers can create a similar study using other methods and approaches.

To conduct a study on the interrelation of the NDVI and climatic variables, one should have all the data available in good condition for the most precise and accurate results. Therefore, it is recommended to sort out all the details and nuances with the data before the research is even begun in order to avoid confusion and mistakes. Simply put, data management is highly important because the more data that is available, the bigger the scope and the scale of the study.

Future researchers are highly recommended to combine the NDVI with other vegetation indices. Owing to the high-level performance of the GEE software program, it is guaranteed that the results much better and more accurate. The GEE is much more advanced in terms of geospatial analysis and image processing, which probably make it the best application to make land use and land cover classification and change detection analysis. Moreover, the GEE catalog contains a lot of different information that can be used for land cover monitoring. However, the spatial resolution of this data is not appropriate for the analysis at the local level. But, the functionality of GEE for users to upload their own data allows for a fulfilling analysis of small areas using more detailed data. Users can capture their own information with an unmanned aerial vehicle (UAV) and use it for better land management. The GEE functionality showed great usage potential for integrated water resources management since it allows for the use of tools for data processing along with the data catalog. Yet there are still some difficulties in creating final maps and reports, because this platform focuses more on data analysis. Thus, some additional GIS software such as QGIS, ArcGIS, or GMT needs to be used.

References

1. Afghan weather information. https://www.weather-atlas.com/en/afghanistan/
2. Catalog of glaciers SSRR, vol 14, Central Asia, Issue 3, Part 5, L. Hydrometeoizdat (1980)
3. Didan K (2015) MOD13Q1 MODIS/Terra vegetation indices 16-day L3 global 250m leSIN grid V006, distributed by NASA EOSDIS land processes DAAC
4. Drusch M, Del Bello U, Carlier S et al (2012) Sentinel-2: ESA's optical high-resolution mission for GMES operational services. Remote Sens Environ 120:25–36. https://doi.org/10.1016/j.rse.2011.11.026
5. FAO (1993) The Islamic Republic of Afghanistan watershed atlas. Food and Agriculture Organization of the United Nations (FAO), Rome
6. Favre R, Kamal GM (2004) Watershed atlas of Afghanistan. Helolse Condroyer, Atlas Communication
7. Gorelick N, Hancher M, Dixon M et al (2017) Google Earth Engine: planetary-scale geospatial analysis for everyone. Remote Sens Environ 202:18–27. https://doi.org/10.1016/j.rse.2017.06.031
8. Huete AR, Didan K, Miura T et al (2002) Overview of the radiometric and biophysical performance of the MODIS vegetation indices. Remote Sens Environ 83:195–213
9. Kalnay E, Kanamitsu M, Kistler R et al (1996) The NCEP/NCAR 40-year reanalysis project. Bull Am Meteorol Soc 77:437–472
10. Lashkaripour GR, Hussaini SA (2007) Water resource management in Kabul river basin, eastern Afghanistan. The Environmentalist 28(3):253–260. https://doi.org/10.1007/s10669-007-9136-2
11. Lobell DB, Thau D, Seifert C et al (2015) A scalable satellite-based crop yield mapper. Remote Sens Environ 164:324–333. https://doi.org/10.1016/J.RSE.2015.04.021
12. Mack TJ, Akbari MA, Ashoor MH, Chornack MP, Coplen TB, Emerson DG, Hubbard BE, Litke DW, Michel RL, Plummer LN, Rezai MT, Senay GB, Verdin JP, Verstraeten IM (2010) Conceptual model of water resources in the Kabul Basin, Afghanistan. In: U.S. geological survey scientific investigations report 2009–5262
13. Pekel J-F, Cottam A, Gorelick N, Belward AS (2016) High-resolution mapping of global surface water and its long-term changes. Nature 540:418
14. Potin P, Rosich B, Miranda N, Grimont P (2016) Sentinel-1 mission status. Procedia Comput Sci 100:1297–1304. https://doi.org/10.1016/j.procs.2016.09.245

15. Sazib N, Mladenova I, Bolten J (2018) Leveraging the Google Earth Engine for drought assessment using global soil moisture data. Remote Sens 10:1265. https://doi.org/10.3390/rs10081265
16. Shelestov A, Lavreniuk M, Kussul N et al (2017) Large scale crop classification using Google Earth Engine platform. In: 2017 IEEE international geoscience and remote sensing symposium (IGARSS), pp 3696–3699
17. Sidhu N, Pebesma E, Câmara G (2018) Using Google Earth Engine to detect land cover change: Singapore as a use case. Eur J Remote Sens 51:486–500. https://doi.org/10.1080/22797254.2018.1451782
18. Tucker C, Pinzon JE, Brown M (1994) Global inventory modeling and mapping studies
19. World Bank (2004) Water resource development in Northern Afghanistan and its implications for Amu Darya Basin. World Bank working paper no. 36. Washington
20. Xiong J, Thenkabail PS, Gumma MK et al (2017) Automated cropland mapping of continental Africa using Google Earth Engine cloud computing. ISPRS J Photogramm Remote Sens 126:225–244. https://doi.org/10.1016/J.ISPRSJPRS.2017.01.019

The manufacturer's authorised representative in the EU is Springer Nature Customer Service Centre GmbH, Europaplatz 3, 69115 Heidelberg, Germany. If you have any concerns regarding our products, please contact ProductSafety@springernature.com

Printed and bound by CPI Group (UK) Ltd, Croydon, CR0 4YY
30/09/2026
02225495-0006